BRAIN TRUST

BRAIN TRUST

The Hidden Connections Between Mad Cow and Misdiagnosed Alzheimer's Disease

Colm A. Kelleher, Ph.D.

Paraview Pocket Books

New York   London   Toronto   Sydney

PARAVIEW
191 Seventh Avenue, New York, NY 10011

POCKET BOOKS, a division of Simon & Schuster, Inc.
1230 Avenue of the Americas, New York, NY 10020

Library of Congress Cataloging-in-Publication Data

Kelleher, Colm.
  Brain trust : the hidden connection between mad cow and misdiagnosed
Alzheimer's disease / Colm Kelleher.
    p. cm.
  Includes bibliographical references and index.
  ISBN: 0-7434-9935-2 (alk. paper)
  1. Prion diseases. 2. Alzheimer's disease. I. Title

RA644.P93K44 2004
616.8'3—dc22   2004053505

First Paraview Pocket Books hardcover edition October 2004

10 9 8 7 6 5 4 3 2 1

POCKET and colophon are registered trademarks of
Simon & Schuster, Inc.

Designed by Jaime Putorti

Manufactured in the United States of America

For information regarding special discounts for bulk purchases,
please contact Simon & Schuster Special Sales at 1-800-456-6798
or business@simonandschuster.com.

To Donal, Maresa, Shane, Devon,
Jaime, David, and Dawn.

To Donal, Maresa, Shane, Devon,
Jaime, David, and Dawn.

Contents

1

The End

During the spring of 2003, in the small town of Goldendale, Washington, the deputy sheriff, a patrolman, and a veterinarian faced something they had never seen before. They had been called to the scene of a gruesome crime. A nine-month-old Gelbvieh bull lay on the ground with its scrotum removed in a neat circle. The bull's penis had been pulled out and excised. The animal's left eye was gone. The tongue had been removed, the cut going deep into the back of the throat. No footprints or tire tracks had been found near the dead bull. Normally when cattle die, the final leg movements can leave marks on the ground. Here the investigators found no signs of struggle. According to the owner, the animal had been healthy the day before. It was obvious to the veterinarian that the nine-month-old animal had been deliberately killed and mutilated. Somebody had used a sharp instrument and some surgical know-how to remove the animal's sex organs. The veterinarian had seen enough to be able to tell the difference between predators, scavengers, and the skilled use of a surgical instrument; the dexterous removal of the tongue gave it away. Either the coyotes were packing scalpels that day or the people who had carried this out knew what they were doing.

A month later, the investigators were still scratching their heads. Three more previously healthy cows had been killed silently in the night on the same property using the trademark lethal skill and a thorough knowledge of surgical procedure. Again, the sex organs and eyes had been removed, and in two cases, the tongues had also been taken out. These were not predator attacks. For a second time, the location of the cow killings and mutilations was Goldendale, a sleepy hamlet that lies on the southern border of the Yakima Native American reservation. The region is not populated.

Six months later, on December 23, 2003, authorities announced that the first case of mad cow disease in the United States had been found in a Holstein dairy cow in nearby Mabton, Washington. Mad cow disease, also known as bovine spongiform encephalopathy (BSE), is a deadly and mysterious disease that afflicts cattle. The infectious organism is known as a prion. It is not a virus and it is not a bacterium. It is a unique infectious protein that kills by provoking suicide programs in billions of nerve cells and leaving a spongy, misshapen mess of brain in its wake. Under a microscope the brain of a BSE victim is full of holes. Typically, a few years after initial infection, cattle begin to stagger and act strangely. They twitch and fall down and their behavior can become bizarrely aggressive. Hence the moniker "mad cow." The disease progresses inexorably and the animals die. There is no cure.

The first announced case of BSE in North America took place just sixty miles from the scene of the unsolved cattle mutilations that had been investigated by law enforcement and a veterinarian a few months earlier. A coincidence? Perhaps. But it was not the first time it had happened.

The story of the discovery of America's first mad cow began on December 9, 2003, the day that a Holstein cow from a Mabton farm was slaughtered. As reported by Steve Mitchell, the UPI medical correspondent who is credited with much of the important investigative journalism on this breaking story, "The cow was chosen for testing because it was a so-called downer, meaning it was unable to walk or stand—a possible sign of the disease." But was it really a downer? The next sentence of Mitchell's story would provide an alternate possibility: "Although in this case, an injury to the pelvis sustained while giving birth appeared to be the cause of the cow's inability to walk." In any case, the Holstein's brain was then shipped to the National Veterinary Services Laboratory, USDA's mad cow testing lab in Ames, Iowa. The lab began to prepare the brain tissue on December 11. The sample was ready for analysis on December 22, after the usual ten days or so the USDA test takes to fix and stain the brain tissue.

The next day, just as people were doing their last-minute Christmas shopping or leaving town to visit relatives, the results of the testing were made public: America had its first case of mad cow disease. "By that time," reported Mitchell, "meat from the animal had been turned into ground beef and already made its way to supermarkets in several different states where some consumers unwittingly purchased and ate it." As the USDA continued to make announcements that corroborated Mitchell's reporting, American meat eaters shuddered. So did the cattle industry; the domestic beef and dairy market is worth tens of billions of dollars in the United States.

Beginning on December 23, the cattle industry canceled all Christmas vacations for their spin machine, and over the holidays began an orchestrated campaign to persuade the American public that eating beef was safe. A key person in the USDA Christmas counterattack was Alisa Harrison, the spokesperson for Ann Veneman, the Secretary of Agriculture. Over the two weeks that followed Harrison proceeded to issue statements, manage press conferences, and otherwise guide the news coverage of the mad cow crisis, all in an effort to reassure "the world that American beef is safe," according to *Fast Food Nation* author Eric Schlosser in an Op-Ed piece in *The New York Times.*

Some people might think that Harrison must have a scientific or veterinary background in order to make such statements. Those people would be wrong. "Before joining the department," noted Schlosser, "Ms. Harrison was director of public relations for the National Cattlemen's Beef Association, the beef industry's largest trade group, where she battled government food safety efforts, criticized Oprah Winfrey for raising health questions about American hamburgers, and sent out press releases with titles like 'Mad Cow Disease Not a Problem in the U.S.' Ms. Harrison may well be a decent and sincere person who feels she has the public's best interest at heart.

Nonetheless, her effortless transition from the cattlemen's lobby to the Agriculture Department is a fine symbol of all that is wrong with America's food safety system. Right now you'd have a hard time finding a federal agency more completely dominated by the industry it was created to regulate."

The top echelon of the USDA is literally packed with former executives from the beef industry. Veneman's chief of staff was previously the chief lobbyist for the cattlemen's association. Other veterans of the association also have high-ranking jobs at the USDA. So do some former meatpacking executives and a former president of the National Pork Producers Council, Schlosser pointed out. Given the seamless revolving door between the USDA and the cattle industry, the obvious question becomes, does the USDA have the interests of the American consumer at heart or does it have the interests of the cattle industry as a higher priority? Are the statements emanating from the USDA to be interpreted as the unvarnished, objective analysis of a government agency that is taxpayer funded and is working on behalf of the people?

The spin machine from the USDA seemed to work like clockwork until Dave Louthan went public a month later with his account of the slaughter. Louthan is the person who killed the mad cow at Vern's Moses Lake Meat Company on the ninth of December. Only Louthan says the cow was not a downer. He says the cow was perfectly healthy-looking, and the only reason she was killed was because she refused to get off the trailer ramp and he feared she might back up and trample the other cows lying prostrate in the trailer. It was late in the day, the cow looked balky, and "I was cutting corners," Louthan told the reporter for the *Seattle Times*. Louthan had been in the job for four years and has no doubt he remembers the right cow: "Every cow that comes in there, I kill. That kind of puts us in a relationship."

The *Seattle Times* story continued: "So he shot a bolt through her head, scooped out a bit of brain, put it in a bag, labeled it with her number, and hung it on the wall with samples from others in the truckload. Later, he checked records to confirm that the 'mad cow' was the cow he remembered, the balky Holstein from the Sunny Dene Ranch in Mabton, Yakima County." If he had not killed the cow outside, the mad cow might never have been discovered, because the plant's testing program called for sampling cows killed outside only.

Before long, others came forward to corroborate Louthan's account. The plant manager, Tom Ellestad, confirmed that the cow was walking. "She did walk off the trailer at our place," Ellestad bluntly told *The Oregonian*. "You got a definition (problem) with the USDA and the veterinarian. They certainly want to call it a downer." The trucker who hauled the animal to the slaughterhouse also said the animal was walking. And an investigation by the Government Accountability Project, a citizens' watchdog group, also confirmed the story by Louthan, who was laid off within weeks of the incident, because, according to his bosses, business had slowed down. Only one person claimed the animal was a downer. The USDA released the actual veterinarian record of the animal; it stated that the animal was a downer.

Every year 35 million cattle are slaughtered in the United States. Out of this number about 200,000 cows are labeled as downers. Until December 2003, the USDA routinely tested

about 20,000 downer animals per year for evidence of mad cow disease. Now they test more, but still only a small fraction of the total number of cattle slaughtered every year. So if the animal, as three witnesses have testified, was not a downer, then there is no chance that the USDA could possibly have picked up a case of mad cow disease. And even then, since they were only testing about 10 percent of all downers, even this level of testing was a mere gesture.

That's why Louthan's testimony sent shock waves through the cattle industry. Their whole policy was based on a gamble that only downer cows would have mad cow disease. Louthan's testimony exposed the charade of this policy by showing that a walker could just as easily have mad cow disease.

Although this shocking discrepancy received some attention from the media, it did not get the front-page coverage it deserved. No one was really aware of the implications of the conflicting stories about what happened outside Vern's meatpacking facility in remote Washington State on that fateful day in December 2003. If Louthan's version of the events is correct, then this case of BSE was caught purely by accident. The heart-stopping implication is that there are probably many more cases out there like this one. And that means the American public may be happily consuming mad cow burgers.

This is a complex, multifaceted story. It is one that I have followed for twenty years as an interested scientist. But only recently did I come to realize that my own research of the past decade actually provides a small piece to this puzzle. As a result, I have gone back and researched the story, found some of it was well known and some of it was buried in the scientific literature. I have also interviewed a number of the principals in this story, which, as we now know, began in New Guinea nearly fifty years ago.

This book suggests how the well-meaning efforts to research a deadly disease that killed thousands of people in New Guinea may have had unforeseen, terrible consequences. The period following these experiments has seen a burgeoning contamination of the North American food chain, and a dramatic increase in the number of people dying from Creutzfeldt-Jakob disease in this country and elsewhere. Joining these dots has, until now, remained below the radar of public consciousness. But on December 23, 2003, all that changed with the public unveiling of the first official case of mad cow disease in a North American herd. Here now is the full story.

2

Kuru

A young woman sat in the corner of the mud hut. Though she was bone thin, it was the look on her face that startled her visitor. The woman's face was expressionless. Her eyes were blank. The lack of expression was so profound, in fact, that she could have been wearing a flesh-colored mask. Every few minutes, a fluttering tremor ran through her body, as if she was shivering uncontrollably from a cold wind.

Vincent Zigas observed the woman as he sweated inside the hot, humid hut. Zigas had never seen symptoms like these before. He learned that the woman had kuru. She had been bewitched, he was told. There were many women and children in the village who had been bewitched by powerful sorcerers and they would all die, or so the story went. Nobody recovered from kuru.

Zigas was a young German-Lithuanian doctor from Australia. He had arrived in the central New Guinea Highlands in 1955 on an Australian government assignment to help eradicate some of the diseases that thrived in the hot, clammy climate. Papua New Guinea presented a daunting topography for outsiders. Within an area slightly larger than California, the country combines dense, almost impenetrable, leech-infested rain forests and highlands stretching up to more than 13,000 feet. Half the island had belonged to Australia since 1910, and after the Japanese takeover during World War II, the Australian government sent several officials into the Highlands in a bid to tame the rampant lawlessness and often gory tribal conflicts among its warlike inhabitants who still used bows, arrows, and stone axes, and knew nothing of the existence of the wheel. But the term "Stone Age" fails to capture the real flavor of the place. Since Captain Cook's forays into the region in the late 1700s, New Guinea had a reputation for being home to multiple tribes of bloodthirsty cannibals and headhunters.

After his arrival in Kainantu and a nearby village called Okapa, Zigas discovered he was the only medically trained doctor in that part of New Guinea. In 1955, Kainantu, known as the "gateway to the Highlands," was a small settlement several days' hike from the Highlands, where the medical supplies from the Australian Department of Health arrived. The town was a central crossing point for people moving up and down from the Highlands and was a natural place for Zigas to set up his base. While in Kainantu he heard rumors about a mysterious disease called kuru in an obscure tribe called the Fore (pronounced FOR-ay). The Fore lived in the remote highlands and had had very little contact with the outside world.

In September 1955, accompanied by a guide, Zigas set off to investigate these increasingly persistent rumors. After two days' hiking in the high terrain the guide led him into a small hamlet with a few scattered mud huts where Zigas witnessed the woman with the strange symptoms. By the end of the year, he had seen dozens of similar cases, mostly in women and children. He first thought it was a brain disorder, maybe a virus or bacterial infection. With almost no medical facilities and no clean water or electricity in the bush, Zigas took what medical supplies he could carry on the two- or three-day hike into the Highlands. As the numbers of kuru cases multiplied, he was quickly overwhelmed.

Kuru was ripping apart the fabric of the Fore tribe, because every death from kuru demanded a death in revenge of the presumed sorcerer who had cursed the victim. The ritual murder, called *tukabu*, usually followed the kuru death by a few days. The deaths from kuru were predominantly women and children, but the deaths from *tukabu* were often men. Usually the *tukabu* involved the unfortunate person who was accused of sorcery, often with no evidence, being bludgeoned with rocks or hacked to death with machetes. A fiendish balance in mortality seemed to be playing out between the deaths of women and children by kuru and the deaths of men by *tukabu*.

In his memoirs Zigas described one of the many tragic kuru cases he witnessed. Walking past a village hut, he had encountered a Fore woman, who held "on her lap a limp figure, grossly emaciated to little more than skin and protruding bone, the shivering skeleton of a boy, looking up at me with blank crossed eyes." Zigas went on to describe how this woman's only child died the next day while her husband had just been murdered in a *tukabu* killing.

Zigas spent a year trying to interest the Australian health authorities in the disease and received only vaguely expressed promises in return. Sir Frank McFarlane Burnet, director of the Walter and Eliza Hall Institute in Melbourne and one of the most famous medical scientists in the world, showed curiosity about the descriptions of kuru but did not assign any of the large number of medical researchers at Walter and Eliza to help the floundering German-Lithuanian. Undeterred, Zigas began gathering samples while working virtually alone in very primitive conditions. By scrounging around among his acquaintances and by begging for funds from the Australian health authorities, Zigas began the long project of building a primitive field hospital near the airstrip at Kainantu.

Zigas impressed everyone with his sincerity and he gradually earned trust among the Fore as well. The reward came in early 1957 when they allowed two tribal members suffering from kuru to make the long trip down to the hospital for medical observation and treatment. The Fore were humoring Zigas. They believed Zigas was wasting his time; they knew that kuru was the result of sorcery and that the only way to cure it was to find the sorcerer and persuade him to lift the curse. So it was that Zigas began the long process of medical detective work with the first two kuru patients at his makeshift medical facility.

Then, on March 14, 1957, a surprise visitor showed up at

Zigas's facility in Kainantu. His now famous description of the caller appears in Zigas's posthumously published book: "At first glance he looked like a hippy, though shorn of beard and long hair, who had rebelled and run off to the Stone Age world. He wore much-worn shorts, an unbuttoned brownish plain shirt revealing a dirty T-shirt, and tattered sneakers. He was tall and lean and one of those whose age was difficult to guess, looking boyish with a soot black crew cut unevenly trimmed as if he had done it himself. He was just plain shabby. He was a well-built man with a remarkably shaped head, curiously piercing eyes and ears that stood out from his head. It gave him the surprised, alert air of taking in all aspects of new subjects with thirst. . . . I guessed him to be from America. . . ."

The caller, the thirty-seven-year-old D. Carleton Gajdusek, had brashly walked in on Zigas as he was preparing to take a trip to the Fore area. By any standards this tall, thin stranger who "machine gunned" people with a constant flow of questions was a remarkable individual. And Gajdusek came to the wilds of New Guinea with some very powerful connections. These connections would have a huge impact on defining the mysterious disease that ailed the Fore.

Gajdusek was a James Bond–like figure capable of slipping into any foreign country, even without permission if necessary. Combining a razor-sharp physician's intellect with fluency in nearly a dozen languages, Gajdusek moved as easily through remote tribes in obscure countries as he did in conversing with the world's best and the brightest researchers in medical science. When he arrived in New Guinea, Gajdusek was well used to spending months sleeping in flea-infested huts and primitive conditions in any number of countries around the world. But it was his very powerful backing in Washington, D.C., that Gajdusek brought to Kainantu that was to change Vincent Zigas's life.

Gajdusek had traveled extensively in South America, the Middle East, and Central and Southeast Asia under U.S. Army funding obtained by Dr. Joseph Smadel. Smadel played "M" to Gajdusek's "James Bond." Gajdusek described the formation of his relationship with Smadel in his Nobel Foundation autobiography: "In 1951 I was drafted to complete my military service from John Enders' laboratory at Harvard to Walter Reed Army Medical Service Graduate School as a young research virologist, to where I was called by Dr. Joseph Smadel. I found that he responded to my over-ambitious projects and outlandish schemes with severity and metered encouragement, teaching me more about the methods of pursuing laboratory and field research, and presenting scientific results, than any further theoretical superstructure, which he assumed I already possessed."

During the 1950s, Joe Smadel was one of the most influential and powerful men in the United States medical establishment. Not only was he director and chairman of the U.S. Armed Forces Commission on Viral and Rickettsial Diseases at Walter Reed Hospital, Smadel was also a central figure in establishing the United States military's embryonic biological warfare program.

A glance at Gajdusek's research activities for Smadel in the year prior to his arrival in New Guinea gives an idea of his ruthless global pursuit of infectious organisms. In a status report dated January 4, 1956, Gajdusek reported on blood samples with antibodies containing poliovirus, herpesvirus, mumps, panleukopenia virus (PLV), and rickettsia from children from the Río Guapay in Bolivia and the Peruvian Amazon. He had conducted seroepidemiology studies of mumps, PLV, toxoplasmosis, leptospirosis, and syphilis throughout Afghanistan, Iran, and Turkey, and had surveyed poliomyelitis and Q fever in the Middle East. He had also collected and dispatched live biological samples of tularemia, and Omsk and Crimean hemorrhagic fevers from the wilds of Central Asia to Smadel's headquarters at Walter Reed Hospital. By overcoming a series of insurmountable obstacles with unorthodox strategies, Gajdusek succeeded in grabbing whatever infectious disease prize he was assigned to capture.

Throughout the 1950s Gajdusek mailed a steady stream of live biological samples back to Smadel. The full title of contract DA-49-007-MD-77 that funded Gajdusek's travels was "Field Studies on the Control of Infectious Disease of Military Importance." In a letter from Smadel to Gajdusek dated December 12, 1955, Smadel wrote: "I want the exact reference to Crimean and Omsk HF agent. I may be able to work a trade for EEE, WEE, and VEE and the two HFs through the microbiological strain center in Lucerne, Switzerland." Translating the acronym-laden scientific jargon, we can see that Smadel was casually trading on a global scale in a large number of infectious organisms of biological warfare importance, including Crimean and Omsk hemorrhagic fever, equine encephalitis virus, Venzuelan equine encephalitis virus, and many others. Thus, by the time Gajdusek's travels took him to remote Papua New Guinea, a successful multiyear relationship between two remarkable men had been established. During the mid-1950s Smadel shifted his operations to the National Institutes of Health (NIH) and maintained a rapidly expanding medical research empire based both in Bethesda and in Camp Detrick, Maryland. Camp (later Fort) Detrick quickly became the center of the U.S. biological warfare research program.

Not only was Gajdusek confident in his abilities and unorthodox methods, he probably also sensed Smadel's interest in the clandestine aspects of Soviet knowledge in the realm of infectious disease as well. "My passport . . . is now clear for the USSR," he wrote to Smadel on December 29, 1956. "My knowledge of the language, literature and places is better than ever. If you can think up any good reason for visiting the USSR en route home (I shall both be under its soft under-belly in Central Asia and on its European borders during the spring) please let me know. I think I would find little difficulty getting around, seeing plenty, learning much, and exchanging ideas. Not being myself aware of anything 'classified' I ought to be a 'safe' exchange ambassador with enough time to spare to learn what wisdom our Soviet colleagues will pour out."

Until his death in 1963, Joseph Smadel would have his finger directly on the pulse of all important infectious disease research conducted by the United States government after World War II. He is perhaps best known for his involvement with the polio vaccine initiative. The drive to eliminate polio from the United States took place just as Carleton Gajdusek was trotting across the globe at Smadel's behest. At that time, in the 1950s, the Sabin and Salk vaccines were in direct competition for effectiveness, and Smadel played a prominent role in judging their efficacy. Publicly, Smadel claimed that the preparations were as safe as Grade "A" pasteurized milk. But privately his sentiments were quite different.

Despite the haggling over which vaccine was superior, both shared a frightening defect not discovered until 1960. In that year, Dr. Bernice Eddy, while working in one of Smadel's many laboratories, found that extracts of the monkey kidney-cell cultures used to grow poliovirus induced malignant tumors in newborn hamsters. When Dr. Eddy had brought up her concerns to Dr. Smadel, he brushed them off, insisting that the cancers were not really cancers but merely "lumps."

But two other virologists, Dr. Benjamin Sweet and Dr. Maurice Hilleman, soon were able to pinpoint a virus, SV 40, as the cancer-causing agent. By this time, millions of children had received polio vaccines contaminated with SV 40 virus, and there was no way of knowing whether medical science had conquered polio at the risk of provoking a new affliction. "Joe Smadel couldn't believe it," recalls a Smadel colleague, Dr. Anthony Morris. "It was a frightening thing. It's still frightening. That information was held up for two years before it was made public, and I saw Joe Smadel fall apart under the pressure of keeping it quiet."

Smadel's other claim to fame was his intense interest in research on mosquitoes as efficient vectors for a family of viruses called flaviruses (including West Nile virus) that caused hemorrhagic fevers in several parts of the world. And it was his interest in arthropod (insect)-borne viruses that in part led him to commission Gajdusek to travel the world collecting samples of mosquito-borne viruses. (It is no small irony that the years 1999–2004 have seen an increase in deaths from West Nile virus in the United States as a result of the spread from mosquitoes.)

So whether it was gathering tissue or blood specimens of tularemia or West Nile virus for biological warfare development, or samples of polio serum from infected children, by 1957 the Smadel-Gajdusek team was a well-oiled machine for obtaining live infectious disease organisms from anywhere in the world.

After their initial meeting in mid-March of 1957, Gajdusek lost no time in accompanying Zigas to visit the two patients in his nearby field hospital. The two women could no longer walk and shivered uncontrollably. Their arms and legs pulsed with slow, continuous, involuntary tremors, which Gajdusek would later call "athetoid movements." Their speech was slurred, their smiles silly, and their grimaces prominent. Gajdusek would coin a phrase for this, too; he called it "pathological laughter."

Gajdusek's first letter to Smadel, dated March 15, 1957, described the excitement he felt after he first saw the mysterious kuru: "I am in one of the most remote, recently opened regions of New Guinea (in the Eastern Highlands) in the center of tribal groups of cannibals, only contacted in the last ten years—still spearing each other as of a few days ago and cooking and feeding the children the body of a kuru case . . . only a few weeks ago. To see whole groups of well-nourished healthy young adults dancing about with athetoid tremors which look far more hysterical than organic is a real sight. And to see them, however regularly, progress to neurological degeneration in three to six months, usually three, and to death is another matter and cannot be shrugged off."

To Zigas's dogged medical investigation skills, Gajdusek brought his capacity to work continuous eighteen-hour days. The team would prove formidable. Within a couple of months, the two physicians had begun equipping the rudimentary field hospital with facilities for taking blood samples and for conducting autopsies. Zigas and Gajdusek spent much of their time trying to treat the steady stream of kuru patients in their hospital. The remainder of the time they spent hiking through the rough terrain, trying to map the epidemiology and geographical distribution of kuru.

Gajdusek accompanied Zigas on several grueling expeditions in 1957. In his book titled *Laughing Death*, Zigas describes Gajdusek's indomitable thirst for kuru research: "Our six day trek provided an opportunity to observe even more pictorial parables of Carleton, the crewcut maverick. His strength and endurance were outstanding. Upon our arrival in a village after the most strenuous 'thrills,' soaked to the skin, numbed and short-winded, Jack and I would have to rest for a while. Carleton, however, would immediately commence to interview the villagers and collect blood specimens. There was a smack of fanaticism in the way he collected blood from every willing person, including infants, regardless of sex or age. Although I had to go along with his 'folly' I did not dare bleed infants or toddlers."

In another display of his endurance, Gajdusek embarked on a 1,500-mile trek through some of the most dangerous travel conditions in the world in order to map the outer epidemiological limits of kuru and determine how far it extended into the Highlands. Gajdusek found that kuru stopped abruptly at the outer perimeter of the Fore domain. It appeared to be exclusively confined to the single tribal group.

In patrolling the outer limits of the kuru epidemic area, Gajdusek noticed two features that did not accord well with a nutritional cause for the disease. "Many cases develop outside the kuru region in Fore or kuru-region people who have moved into the kuru-free populace to live and work," wrote Gajdusek. "The illness progresses relentlessly even if patients are totally removed in earliest stages from the kuru region and are on a diet of the non-kuru regions." He then went on: ". . . it means that a most unusual toxin exposure is involved, one in which exposure months or even years previously is sufficient to initiate a progressive slow neurological destruction." Gajdusek was beginning to flirt with the notion of a "slow virus" as the cause of kuru.

Gajdusek's work in New Guinea continued at a frenzied pace, as did his lengthy and detailed missives to Smadel. In a letter from the remote Okapa police patrol post in New Guinea dated April 3, 1957, Gajdusek wrote: "We have now located forty-one cases of kuru, a clear cut central system degenerative disease of rather rapid course and so uniformly and progressively devastating and almost always fatal . . . we know of another forty cases in addition to our current forty one, who have had the disease and died . . . We have accumulated histories of well over a hundred other cases . . ."

Almost from the first days, Gajdusek's intense personality and his obvious zeal in probing the mysteries of kuru captivated the Fore people. The Fore respected his commitment to searching out new cases of kuru, even though they knew all attempts to solve the mystery were folly because "kuru was sorcery" and was immune to Western medicine. His superb linguistic skills were revealed early on and he quickly learned the basics of communication with the tribe. So it was not surprising that midway through 1957 Gajdusek had persuaded the Fore to allow them to cut into a kuru victim's brain after death.

Meanwhile, within a few weeks of Gajdusek's arrival, political tensions had exploded. McFarlane Burnet and the medical establishment at Walter and Eliza Hall Institute at Melbourne began to regret their casual attitude in not supporting the multiple pleas for help from Vincent Zigas over the previous eighteen months. Suddenly, within a few months, an American with very powerful connections had thrust himself into the middle of "their research" and threatened to hijack any fame and fortune that might result from the discovery of a brand-new neurological disease.

Almost immediately, the Australians began to send supplies and personnel to Kainantu to counteract what they saw as an attempt by the Americans to usurp their research. They also began to apply enormous pressure to protect any biological samples that were obtained from being sent to the United States. When McFarlane Burnet discovered that Gajdusek was about to perform a brain autopsy on a Fore woman who had just died of kuru, he insisted that the brain be shipped to Melbourne.

With typical derring-do, Gajdusek described this first autopsy in a letter to Smadel: "I write at the moment to tell you that we have had a kuru death with a complete autopsy. I did it at 2 AM during a howling storm, in a native hut by lantern light, and sectioned the brain without a brain knife."

3

First Link

The first kuru brain went to Melbourne, but Gajdusek had promised Smadel the second and he kept his word. Within a short time, a middle-aged woman called Yabaiotu had succumbed to kuru and this time Gajdusek sent the carefully autopsied brain back to the United States. In the intervening couple of months, Smadel had hired an expert neuropathologist named Igor Klatzo to examine Gajdusek's exports at the NIH labs. During the summer of 1957, Gajdusek sent five more brains to Klatzo.

It did not take Klatzo long to report back. By October of 1957, Klatzo had some revolutionary findings on hand. He reported extensive damage to Yabaiotu's cerebellum as well as to other brain structures. The damage did not fit the profile of any disease he had encountered. Klatzo initially believed that some toxic substance in the Fore diet or maybe in their environment was responsible. Gajdusek had spent months

tracking Fore diets in exquisite detail and at one time he suspected heavy metal toxicity.

Klatzo also reported that the damage to the six kuru
brains seemed to resemble the lesions found in Creutzfeldt-
Jakob disease. When Zigas read Klatzo's diagnosis, he
scratched his head. But Gajdusek had heard of Creutzfeldt-
Jakob disease (CJD), and he lost no time in explaining the
disease to Zigas.

Hans Gerhard Creutzfeldt lived in Breslau (now Wrocław),
Poland, at the beginning of the twentieth century. He
worked in the clinic of another physician whose name was
destined to be uttered with dread: Alois Alzheimer. At that
time Alzheimer was a well-known neuropathologist with a
reputation that spanned Europe.

On June 13, 1913, twenty-three-year-old Bertha Elschker
was brought into the clinic at the University of Breslau. During Creutzfeldt's interview with the new patient, she
recounted some puzzling symptoms. "In May 1913," wrote
Creutzfeldt, "the patient . . . began to walk unsteadily; in
addition to this a mental change appeared. She no longer
wanted to eat or to bathe, she neglected her appearance,
became dirty, complained of pressure in the region of the
heart, assumed peculiar postures, in that she bent over to her
left and pressed her hand against her heart. The unsteadiness
of gait increased rapidly, and fourteen days before admission
the patient fell over while standing, without losing consciousness. There was no evidence of fever. Three days before
admission she suddenly screamed out that her sister was
dead, that she was to blame, that she was possessed of the

devil, that she herself was dead, that she wanted to sacrifice herself." During the preliminary exam, Creutzfeldt noted that sometimes she exhibited spasmodic movements in her facial muscles, alternately grinning and grimacing.

Creutzfeldt admitted Bertha to the neurological clinic at the University of Breslau and began observing her. His detailed description of her condition is noteworthy for the number of different variations as the disease progressed: "During the period of observation the behavior is found to be very variable. At times the patient acts in a silly way, with a tendency to jocularity; she makes word associations in a distractible, entirely superficial manner . . . Often she appears distracted, makes all sorts of grimaces, speaks in an odd, stilted fashion as if she wanted to emphasize even more the scanning quality of her speech. Frequently, there are unmotivated outbursts of laughter, which give the impression of being purely a motor activity. Her attention lapses rapidly, even when momentarily she is aroused to greater alertness . . . The muscular jerking twichings must often be characterized as more than pseudospontaneous. At times these irritative phenomena are pronounced in the upper extremities and face. . . ."

Within a few weeks Creutzfeldt noticed she could no longer walk. Lengthy bouts of humorless laughter would suddenly erupt. Day by day, he noticed, she began to lose recognition of her surroundings and of the people who attended her. All the while she was losing weight because she refused to eat. "On August 6 a genuine epileptic attack occurs, beginning with clonic jerkings in the right arm, then involving the right half of her face; on the left relatively less contractions in the musculature of the shoulder, chest and face are the only signs of involvement; towards evening a second attack, exactly similar to the first." Within a few days, her condition had worsened dramatically. "In the last hours stupor deepens," his notes on the patient concluded, "swallowing is impaired, death ensues on August 11."

In his 1920 paper Creutzfeldt described unusual lesions in her brain. Massive cell death appeared to have occurred throughout the young woman's brain, leaving holes in the gray matter. Repeatedly throughout his description Creutzfeldt emphasized that he could not find any evidence of inflammation; he described "a noninflammatory focal disintegration at the neural tissue of the cerebral cortex," and later in the paper "a noninflammatory diffuse cell disease with cell outfall throughout almost the entire gray substance."

Creutzfeldt's inability in 1913 to find any sign of inflammation in his young patient was important and was a grim foreshadowing of Carleton Gajdusek's failure, forty-four years later, to find evidence of fever or inflammation in the Fore patients who were dying of kuru. One of the signs of infection by a bacterium or virus is inflammation. The inflammation is the response of the immune system to an outside invader and the lack of fever or any other sign of viral or bacterial attack led to great confusion on Gajdusek's part in determining the origin and cause of the disease.

Within a couple of years of Dr. Creutzfeldt's paper another of Dr. Alois Alzheimer's protégés, Dr. Alfons Maria Jakob in Hamburg, began studying two men and three women, ages thirty-four to fifty-one. All five people appeared to have exhibited similar symptoms to those of the unfortunate Bertha Elschker in Breslau—personality changes, loss of memory, loss of the ability to talk, to stand, or to walk manifested at a slow but relentless pace. As in the original case the subjects lost recognition of people and surroundings, and at different stages they all became bedridden. A couple died within a few weeks; the most robust lasted nearly a year after being bedridden. But all five succumbed to the same ruthless, inevitable march toward death. Dr. Jakob wrote three papers, published in 1921 and in 1923, describing the five patients.

Apart from a notable mention of Creutzfeldt-Jakob disease by a colleague who had worked with Dr. Jakob, the early descriptions of these obscure conditions were relegated to the far reaches of academic library shelves. No notable advances in determining the origins of the disease in humans would be made for years to come.

By the time Klatzo noticed in 1957 just how similar the kuru lesions were to those in Creutzfeldt-Jakob disease (CJD), only about twenty cases of CJD had been described, but none had appeared in any English-language scientific journal. Additionally, it was known that CJD affected middle-aged people, not children as in kuru. Furthermore, the kuru lesions appeared to be much more pronounced than those in CJD.

In about half of the kuru cases Klatzo examined, he noticed unusual "plaques" in the brain sections from the cerebellum. Under the microscope, the plaques looked like a bunch of large, hairy fibers organized around dark centers. The plaques reminded Klatzo of larger versions of those seen in Alzheimer's disease. At that time, nobody had described plaques in CJD. When Gajdusek saw the Klatzo findings of CJD-like as well as Alzheimer's-like damage in the cerebellum, he was even more puzzled, since both CJD and Alzheimer's disease were rare conditions. Why were obscure syndromes that affected one in 10 million Europeans killing thousands of Fore tribespeople halfway around the world?

While Zigas, on orders from his superiors in Australia, finished setting up the medical facilities in Kainantu, Gajdusek continued the research field trips into the Highlands virtually on his own. Meanwhile, Gajdusek's letters back to Smadel described his frustration as he threw every known antibiotic, anti-inflammatory, and dietary modification he knew at the Fore with no effect on the disease. Both Zigas and Gajdusek initially assumed kuru was the result of an infection, but no matter how hard they looked, they found none of the telltale inflammation of lymph nodes and no fevers. This disease defied all attempts at treatment.

In November 1957 Gajdusek and Zigas penned two articles that informed the medical world about a brand-new neuropathological disease epidemic. The articles described the clinical picture as well as the preliminary epidemiology of kuru.

By this time, the Australian press began to worship Gajdusek. His outsize personality and his James Bond–like exploits made him a natural focus for reporters. Gajdusek complained to Smadel: *"Time* (Magazine) stopped in to get a kuru-story while I was on the last 'cross-the-island' patrol. Furthermore, in Moresby I was besieged by reporters—even equipped with tape recorders—and I am most disturbed and

annoyed. 'Laughing Death' . . . which is a hideous mis-
nomer, fills most Australian papers and magazines with
highly distorted and well padded accounts. I . . . wish to hell
that kuru were less a thing to fire the popular imagination,
for it will, I fear, be played for all it is worth by the press."

Even the hard-bitten Smadel was thoroughly impressed
by Gajdusek's devotion and his accomplishments. In 1958 he
used his considerable influence to create a position for Gajdusek as a neuropathology researcher at the National Institute of Neurological Disorders and Stroke at the NIH.
Smadel's letter of recommendation included an oft-quoted
description of Gajdusek as "one of the unique individuals in
medicine who combines the intelligence of a near genius
with the adventurous spirit of a privateer."

When Gajdusek returned to Washington in 1958, he
quickly began organizing a traveling kuru exhibit in order to
spread the word of this unique new neurological disease in
humans. The exhibit consisted of large colored photographs
of Fore people, photographs of various victims in different
phases of the disease, and multiple, large, color photomicrographs showing the extensive deterioration in different
regions of the autopsied brains.

One of the visitors to Gajdusek's traveling exhibit would
provide the second key to the puzzle.

4

Mad Sheep

William Hadlow stood spellbound in front of the kuru exhibit in London. Hadlow was a pathologist sent by the U.S. Department of Agriculture to conduct research on scrapie, a usually fatal disease of sheep and goats, at the Compton Laboratory at the Institute for Research on Animal Diseases in England. In one of the curious coincidences that change people's lives, Hadlow's work at Compton was to take an unexpected twist.

Today Hadlow is eighty-one years old. He is as sharp as a tack and lives in remote Hamilton, Montana, next to the Rocky Mountain Labs where he has spent a good part of his life. Hadlow described the coincidence to me, recalling with ease events that happened a half century ago.

"A friend of mine, a colleague named Bill Jellison, was passing through from a conference and stopped by Compton in June 1959," recalled Hadlow. "At dinner that evening he mentioned I might be interested in an exhibit he saw the previous day at the Wellcome Medical Museum in London. It had to do with a strange brain disease of a primitive people in New Guinea called kuru. Soon after Jellison left I took a train into London to see the display for myself. . . . The exhibit was Carleton Gajdusek's kuru show and was made out like a poster session with lots of colored photographs. . . ."

Hadlow was riveted by what he saw. "I was immediately drawn to the lesions in the colored photomicrographs of kuru brains," he told me with a slight inflection in his voice as he remembered the impact of those photographs. "I am sure they were the same colored photos taken by Dr. Klatzo. It immediately struck me right then: Gosh, here is something." Hadlow remembers looking at the photomicrographs showing the vacuoles, or holes, in the neurons, the cells that are the central nervous system's primary means of communication. Holes in neurons are very uncommon in human brains and a sure sign of a big problem. "I had seen them for the past year because I had been looking at scrapie brains from goats and sheep," he recalled. But that was not the only similarity with scrapie. Around the vacuolated neurons were astrocytes, star-shaped cells that accumulate around, and help protect, injured neurons. He had seen these in scrapie brains as well.

At that moment, Hadlow was in possession of knowledge that nobody else had. Not Gajdusek at NIH, nor Zigas in New Guinea, and not Smadel in Bethesda/Fort Detrick. He had seen the link between scrapie and kuru.

It is not clear exactly when or where scrapie first appeared. A recent review in the *British Medical Journal* *(BMJ)* mentions "a suggestion that it was already present in northern Europe and Austro-Hungary before the beginning of the 18th century." In any case, by the middle of the century, the disease had begun to decimate flocks of sheep in Britain. This was a time when there were about as many sheep as people (10 million) in the country, and when wool was one of the dominant commodities in the British economy, accounting for the employment of nearly a quarter of the population.

By 1755 the sheep farmers of Lincolnshire were angry. So much so that they wrote a letter to the British House of Commons stating that a disease they called "rickets" or "shaking" had been decimating their flocks for about ten years. The farmers wrote that this disease was incurable and was "in the blood" of their sheep a few years before the sheep showed signs. They loudly complained to the government about a group of "jobbers" who bought and sold sheep and who indiscriminately mixed healthy sheep with rickets-infected sheep in an attempt to maximize their profits.

One of the best-known and earliest descriptions of the symptoms in sheep came from the Reverend Thomas Comber in England in 1772, who was writing about a disease called "rickets" at that time: "The principal symptom of the first stage of this distemper is a kind of lightheadedness which makes the affected sheep appear much wilder than usual when his master or shepherd as well as a stranger approaches him. He bounces up from his laire and runs to a distance as though he were pursued by dogs. . . . In the second stage of distemper the principal symptom of the sheep is his rubbing against trees, posts with such fury as to pull off his wool and tear his flesh. The distressed animal has now a violent itching in his skin but it does not appear that there is ever any cutaneous eruption. The third and last stage of this dreadful malady seems to be only the progress of dissolution after an unfavourable crisis. The poor animal as condemned by nature, appears stupid, separates from the flock, walks irregularly, generally lies and eats little. These symptoms increase in degree till death which follows a general consumption."

Outbreaks in France were described in the late eighteenth century with the animals suffering from *tremblante* (the shaking), *la vertige* (dizziness) or *la maladie folie* (mad disease). The word *scrapie* describes the incessant and furious rubbing against any rough surface and seems to have been widely adopted by the late nineteenth century, at least in the United Kingdom. By that time scrapie had been described in several countries in Europe. Throughout this period, there was a great deal of confusion about the different symptoms of the disease as it progressed, and many farmers and veterinarians assumed that there were several diseases affecting their flocks.

Although at times scrapie appeared to affect up to 20 percent of some sheep flocks, a truly enormous rate of attrition, the British and Europeans never seemed worried about transmission to humans. After all, they had been eating lamb and mutton for two hundred years. However, during the 1700s, sanitary systems were nonexistent, dead bodies were usually placed in open graves, and disease was rampant. In 1750 in England and Wales the average life expectancy was less than forty years of age. By 1900, that figure had climbed to about fifty. It could therefore be argued that death from scrapie infection was simply masked by the low life expectancy and by the high levels of other diseases. As the twentieth century progressed, the consumption of sheep in Britain decreased and life expectancy increased.

The scientific revolution during the nineteenth century meant that the introduction of scientific principles and scientific equipment became increasingly commonplace and were applied to the problem at hand. In 1898, Charles Besnoit of the École Vétérinaire d'Alfort learned of a mysterious disease that was killing up to one-fifth of the flocks of sheep in southwestern France. For the first time in recorded history, somebody with scientific training and equipment began to study scrapie. Besnoit used a microscope to examine the organs of sheep that had died of the disease: "Upon microscope examination . . . very obvious nervous system lesions were seen: these were located in the spinal cord and in the peripheral nerves."

Besnoit's observations were the first direct confirmation that scrapie was a neurological disease. Large vacuoles and gaps in the nervous system tissue, especially the brain, began to be incorporated as the definitive diagnosis and the unique feature that distinguished it from the legion of other diseases that killed sheep.

But all of Besnoit's subsequent heroic attempts to delineate how the disease was transmitted, or what caused it, failed. He could find no trace of bacteria in attempts to culture the mysterious scrapie agent and sheep that were inoculated with a couple of liters of blood from a sick animal remained healthy even after nine months. Besnoit tried housing two infected sheep with a healthy animal to determine if the agent was transmissible and again saw no evidence that the healthy sheep became ill. He looked at changing the diet of animals to see if what they ate contributed to the mysterious outbreaks and again he came up empty-handed. Although his many experimental trials ended with frustration, Besnoit's landmark contribution was the determination that scrapie was in fact caused by something that destroyed the very fabric of the brain.

Across the English Channel, veterinary science was flourishing and nowhere better than in Edinburgh, Scotland, where the Royal School of Veterinary Studies was gaining reputation as a preeminent college in teaching the veterinarian arts. "The most famous of its graduates," states the school's online history, "was John McFadyean, who entered the college at its low point in 1874 and later became a member of its staff. He then went on to become the principal to the London Veterinary School and has been justifiably claimed as the founder of modern veterinary science in [Britain]. Perhaps his most outstanding moment was his courageous, but so polite, public rebuttal, at a conference in Edinburgh in 1901, of Robert Koch's vehemently held conviction that bovine tuberculosis was little hazard to man." It was only natural at some point in his career that the preeminent British veterinarian would turn his attention to the dreaded sheep disease.

In 1918, McFadyean published a landmark paper on scrapie. His paper was written as a review with pages devoted to lengthy descriptions of the condition as well as detailed explanations of the epidemiology of the disease. He provided the best description of why scrapie received its name: "The primary cutaneous symptom is an itchiness of the skin, manifested by the animal's inclination to rub itself against walls, posts or other solid objects. This itchiness appears to affect the whole integument, as is shown by the fact that scratching of any part of it with the fingers, even at an early stage of the disease, will cause the animal to evince signs of gratification. In consequence of the rubbing movements, the fleece especially along the sides or above the root of the tail, soon becomes more or less ruffled, and by this sign shepherds are usually first led to suspect an animal. The itching rapidly increases in intensity and extends to the legs and head. To relieve the irritation the hind feet are used to scratch the head, and with the same object the animal gnaws or bites its legs. Eventually the diseased sheep is reduced to the pitiable condition of having to spend nearly the whole of its time in scratching, rubbing, or biting in an obviously futile attempt to allay the constant tormenting itch from which it suffers."

Rather than attempting to place sick sheep beside healthy sheep, as Dr. Besnoit had tried in France, the British veterinarian had decided to look at the spread of the disease in the wild over several years to determine if a flock of sheep that were infected with scrapie could pass it on to a neighboring flock. McFadyean named the three farmers he studied Farmer A, Farmer B, and Farmer X. Farmer A had never had a case of scrapie in eleven years. In late 1907, Farmer A bought more than a hundred ewes from Farmer X at an auction. Much later it was determined that Farmer X's flock was infected with scrapie. In February 1909, the first symptoms of scrapie appeared in Mr. A's flock and, all told, thirty of the animals that Mr. A had purchased from Mr. X succumbed to scrapie.

Being an intelligent farmer, Mr. A then got rid of all the animals he had purchased. For a couple of years, Mr. A

thought he had got rid of the dreaded disease, but then the
calamity happened. Less than a couple of years later, scrapie
appeared in Mr. A's original flock. At first it was confined to
a couple of animals, but as the months passed more animals
succumbed. From this powerful evidence of the spread of
scrapie among sheep, McFadyean reported that the disease
was contagious, but that it had an extremely long latency
period, on the order of a couple of years or more. McFadyean's
1918 paper explained why scrapie had never been seen in a
sheep younger than two years of age.

Like Besnoit, McFadyean tried several times to experimentally transmit the disease. His 1918 paper described
multiple attempts to inoculate healthy sheep with varying
amounts of blood from diseased sheep, to inoculate the animals with samples of cerebrospinal fluid, to inoculate them
with "materials from diseased skin," to administer intestinal
contents orally, and to perform numerous close-contact
experiments. McFadyean observed the animals for up to eighteen months and then gave up, selling the inoculated animals
to a butcher. All experimental attempts to transmit scrapie to
healthy sheep were unsuccessful in McFadyean's hands.

Sometimes it's better not to know what can't be done. In
1934, ignoring almost two hundred years of failed attempts
to study scrapie, Jean Cuillé and Paul-Louis Chelle, veterinarians at the University of Narbonne in France, began their own
inoculation experiments. They selected sheep with advanced
scrapie and chose only brain and spinal cord material. Adding
small amounts of sterile salt water, they used a mortar and
pestle to grind the brain and spinal cord into a slurry.

The veterinarians then selected several animals from a

disease-free flock from Narbonne and injected nine with different amounts of the brain slurry. The experiment was severely compromised when seven of the animals had to be put down for other reasons. Finding no disease symptoms in seven sheep several months after the experiment began would have prompted lesser men to walk away. But Cuillé and Chelle kept the two remaining sheep under observation.

Their decision would prove fortuitous. One of the sheep that had been injected with the brain slurry on July 9, 1934, remained symptom free until late September 1935, when the ewe began to exhibit signs of restlessness and appeared suddenly startled and shy for no apparent reason. Two weeks later she began to stagger and soon the tremors began. When they killed the ewe in October, they found the characteristic brain lesions of scrapie. The second ewe lasted twenty-two months before symptoms appeared.

Doctors Cuillé and Chelle had made a major breakthrough. They showed the disease was indeed transmissible and that it had an unprecedented long latency period. This very long delay explained the previous two hundred years of failed inoculation experiments. Like many breakthroughs, their experiments initially were dismissed; others could not replicate the success. In 1939, however, Cuillé and Chelle put all doubts to rest by taking the slurry of brain tissue obtained from one of the ewes that had finally shown symptoms and injecting the material into another healthy ewe. The second-generation transmission experiments were successful. Within a year the ewe showed the characteristic symptoms and her brain showed the ravages of scrapie. The pair of talented veterinarians also showed it was possible to transmit scrapie to goats using brain slurry from an infected sheep.

Finally, in a coup de grâce, Cuillé and Chelle showed that the scrapie agent was small enough to pass through a filter and successfully infect a sheep. The pair of veterinarians believed strongly that the scrapie infectious particle belonged to the newly described family of "filterable viruses" made famous by Peyton Rous of the Rockefeller Institute when he showed that a filterable virus, later known as Rous sarcoma virus, could transmit cancer to chickens. Researchers began to think of the scrapie agent as a "slow virus."

Meanwhile, a much bigger—and quite unintentional—demonstration of the power of inoculated scrapie was taking place across the channel in England. At the Compton Laboratory at England's Institute for Research on Animal Diseases, one of the more prestigious animal research labs in England, director Bill Gordon began a large-scale attempt in the 1930s to create a vaccine against louping ill, a viral disease that was epidemic in sheep. Louping ill was often mistaken for scrapie because the sheep adopted a peculiar looping gait when they walked. In the previous decade, louping ill was shown to be caused by a "filterable virus."

Gordon created a vaccine by obtaining brain and spinal cord tissues from a flock of sheep that had louping ill. He used standard procedures for killing the louping ill virus in the tissue samples by treating them with formaldehyde and alcohol. This procedure killed the virus and made it safe to inject other sheep with the inactivated louping ill virus to produce an immune response. Gordon and his team at Compton in 1936 produced over forty thousand doses of louping ill vaccine and used them in a large-scale vaccination program.

Initially, the louping ill vaccine program was pronounced a spectacular success, but then in 1937 and 1938 sheep began to show signs of scrapie. In 1937 and 1938 alone, hundreds of vaccinated sheep died of scrapie. Gordon was horrified. When he launched an aggressive investigation into how exactly the vaccine had become contaminated, he found a chilling answer to his question. The sheep had somehow contracted scrapie simply by grazing on a pasture that a scrapie-infected flock had grazed on several months before.

Gordon's large-scale inoculation experiment not only demonstrated that scrapie could indeed be transmitted by injection, it also showed that the infectious agent was tough enough to survive prolonged exposure to formaldehyde. Even in the 1930s it was known that formaldehyde could kill most viruses and bacteria. To Gordon the implications were clear. The infectious agent not only endured the extremely harsh formalin treatment, but it also apparently was capable of surviving in the soil for months. No other infectious particle known was capable of surviving both these conditions. The mystery deepened.

The trauma of the louping ill vaccination catastrophe had the positive outcome that the Compton Research Station began to take an active interest in scrapie research and work began in 1940 to try to create a vaccine against the infectious agent. Though the work largely paused during World War II, it picked up again after the war. Gordon's goal was to develop a vaccine against scrapie, just as he had done with louping ill.

The main problem with scrapie was that nobody had any idea as to what was causing the dreaded disease. Was it a virus or bacterium? Cuillé and Chelle in France strongly suspected a "filterable virus." But that explanation had several problems. If this was a virus, it invaded the body without causing the immune system to create antibodies. No antibodies to scrapie had ever been detected in sheep. In addition, all known viruses were inactivated by formaldehyde and most (but not all) infectious organisms did not survive in pastures for months or years. (Some bacterial spores like anthrax and bacillus subtilis were known to survive in soil.) Most infectious diseases from viruses progress much more rapidly than scrapie. After the average virus infection, an easily measurable acute phase is apparent within days or weeks with detectable antibodies. Even during the subsequent slow or latent phase of most viral infections, blood sampling can detect the presence of the virus. Not so with scrapie.

The only definitive way of confirming scrapie infection was to kill the sheep and conduct an autopsy on the animal. The holes in the brain cells, or neurons, were then visible under the microscope. It was apparent that the disease ravaged the nervous system of sheep and seemed to concentrate there, but it was unknown whether the agent was located in other tissues of the animal. In short, if a virus caused scrapie, it was a remarkable virus that broke all the rules.

Gordon began to apply for a series of grants from the British government's Agricultural Research Council (ARC) but quickly ran into bureaucratic obstacles. For political reasons the ARC did not want a high-profile research study conducted on scrapie because it was reluctant to highlight the economic devastation caused by the disease. Increasingly the farming community in Britain mirrored this attitude. Farmers began to realize that it was economic suicide to report scrapie infection because their flock would be flagged and it was difficult to sell any sheep from an infected flock. An unofficial campaign, on the part of the Ministry of Agriculture, began to hide the extent of scrapie's ravages for farmers and to downplay the importance of the problem. Bill Gordon found it extraordinarily difficult to persuade the ARC to fund his scrapie studies.

In response to the stonewalling from the ARC, Gordon took the unusual step of applying for funds in the United States. His strategy was simple because in 1947 scrapie turned up in the United States as a result of the importation of British sheep. The U.S. Department of Agriculture (USDA) suddenly became eager to examine this new disease and Gordon's funding request was approved. His success in sidestepping the ARC didn't amuse the British Ministry of Agriculture.

Gordon still had to obtain permission from the ministry to actually purchase sheep to inoculate with scrapie. The ARC refused to grant him permission. But Gordon was a stubborn Scot who combined a fierce determination with a low opinion of government bureaucrats. He had a personal commitment to researching the scrapie enigma because of his catastrophic experience with the louping ill vaccine. He ran the Compton Research Station as if it was his personal empire and he was not going to allow nameless bureaucrats at ARC to get in the way of scientific progress.

Gordon defied the ARC order and purchased about forty

sheep for his experiments. In the following months, on two separate occasions, Gordon defied direct orders from the ARC to sell his flock of sheep and abandon his experiments. The ARC was then in the position of having to "put up or shut up," and they chose the latter. But for years afterward many scientists at Compton complained about the vindictiveness on the part of the ARC in deliberately starving the research center of funds in an alleged vendetta against Gordon's defiance. Remarkably, in spite of the intransigence of the ARC, the work on scrapie continued to flourish at Compton.

Overseen by the mercurial Gordon at Compton, researchers Iain Pattison and Geoffrey Millson began a series of systematic inoculation experiments. One of their first series of experiments was to determine if scrapie could be transmitted to goats by mashing brains from infected sheep and injecting the slurry into goat brains. They found rapid success and showed the average incubation time was only six months in goats. They also found that scrapie was almost 100 percent transmissible from goat to goat. Both factors were important determinants in increasing the pace of scrapie research. Pattison and Millson also reported low levels of infectious scrapie in the muscles of goats. This was the first troubling demonstration that the agent was not just confined to the brain and spinal cord. The British Ministry of Agriculture, however, consistently downplayed this important finding.

Another significant result during this tremendous burst of research coming from the Pattison-Millson team was that goats appeared to exhibit two distinct sets of symptoms when injected with scrapie. One group they called "drowsy," the second they labeled "scratching." The latter group went through the

well-publicized scratching symptoms before succumbing to
the other symptoms. The drowsy group simply exhibited brain
damage symptoms without going through the classic scratching phase. Importantly, both clinical subtypes were separately
transmissible to other animals in a reliable manner. This meant
that there were two "strains" of scrapie.

Meanwhile, back in the United States, the appearance of
scrapie on a Michigan farm in 1947 moved the USDA to take
notice. When the disease subsequently appeared in California
in 1952 and in Ohio in 1954, the USDA decided to take
action. (By 1950, scrapie had also been reported in Australia
and in Canada.) All of the outbreaks in the United States
were traced to British importations, so the USDA moved to
ban British sheep imports. Agriculture officials searched for
and found research funding, but they were surprised to discover that no research facilities or research personnel on
scrapie existed in the United States. All the research expertise
resided across the Atlantic at the Compton Research Station
and at the Moredun Research Institute in Edinburgh, Scotland. So the USDA decided to send USDA pathologist
William Hadlow to Compton in 1958. Hadlow quickly
began to make strides in scrapie research at Compton because
he had one of the best research teams in the world to work
with. Within a few months Hadlow had expertly broadened
and sharpened the contours of the widespread damage that
scrapie inflicted on the brain. And within a year he was
standing spellbound in front of the kuru posters in Gajdusek's traveling exhibit in London.

"On the way home from the exhibition I stopped at the

Royal Society of Medicine Library and I picked up a few of
the referenced articles I had seen on the poster," Hadlow
recalled. "They confirmed what I already knew. Scrapie and
kuru were the same lesions." The more Hadlow read of kuru's
clinical details and especially of the dramatic brain lesions,
the more certain he became. Kuru and scrapie were most
likely caused by a similar agent.

Hadlow realized that it would do an injustice to science to
sit on such a breakthrough, so within a couple of weeks he
had penned a short letter to the prestigious medical journal
The Lancet. But because of a printers' strike in London, Hadlow feared that his paper might suffer a delay in publication
at *The Lancet*, so he mailed a copy of his submitted letter
directly to Gajdusek as well.

Hadlow's letter to the *The Lancet* dated September 5,
1959, would be a turning point in the unraveling of this
mysterious disease. In it he laid out the many similarities
between scrapie and kuru, then ended his letter with an elec-
trifying suggestion: "It might be profitable, in view of veterinary experience with scrapie, to examine the possibility of
experimental induction of kuru in a laboratory primate, for
one might surmise that the pathogenic mechanisms involved
in scrapie—however unusual they may be—are unlikely to
be unique in the province of animal pathology." Hadlow's
words were to set in motion a train of events that would eventually reveal the ravages of these diseases striking the very
heart of North American life.

5

Breakthrough

When Hadlow's letter in *The Lancet* reached Gajdusek in New Guinea, it must have shocked him. Not only had he completely missed the link between scrapie and kuru, he had never even heard of scrapie. Even though Gajdusek had combed the veterinary journals of the world, he had fixed on an obscure work on *Phalaris tuberosa* toxicity in cattle that "produces a staggers syndrome in sheep which is as close to kuru as a sheep can get I suppose." Gajdusek had begged Smadel to comb the world's literature for papers on *Phalaris tuberosa* toxicity. Gajdusek's focus on animal diseases was probably confined to looking for toxic substances. He had compiled an exhaustive inventory of the plants, animals, and insects that the Fore ate, trying to track down dietary factors, particularly trace metals like manganese, copper, and zinc, but without any obvious evidence that there was a connection to kuru.

Hadlow's bombshell forced Gajdusek to completely rethink the frantic epidemiological and clinical research that he and Vincent Zigas had carried out for two years in the Highlands of New Guinea, a Herculean effort that had failed to make a dent in kuru. Now, Hadlow's work impelled a radically new direction for kuru research. Gajdusek would have to return to the United States and create an inoculation program that would span several years. As an impatient virologist who was used to working with acute infections that lasted for weeks, it must have seemed a daunting task for Gajdusek to contemplate injecting animals with kuru brains and then waiting around for years to see if they came down with kuru symptoms.

Within two months, Hadlow would meet Gajdusek in the United States. In November 1959, Hadlow, along with Bill Gordon, who had been Hadlow's boss at Compton, John Stamp, who was leading the research drive at the Moredun Institute in Edinburgh, and two USDA researchers were on a national tour around the United States trying to explain to sheep farmers why the USDA had begun killing their sheep. The farmers were outraged and very angry. The USDA had decided to bring in the top scrapie experts in the world and try to give U.S. sheep farmers the facts about the disease. Though scrapie was relatively new, it appeared to be spreading quickly.

It was during a speech Hadlow gave at a USDA facility in Washington that the encounter took place. "I saw this crew cut young fellow standing at the back of the room," Hadlow recalled when I interviewed him, "and after my talk he came up to me." Gajdusek literally buttonholed Hadlow at the

USDA lecture series and pressed him for all he knew about scrapie. It was the beginning of a lifelong friendship.

Within a year of meeting Hadlow, Gajdusek would visit both Compton and Moredun to see firsthand the research on the sheep disease that he had so completely missed. Compton was the center of the universe of scrapie research. Compton scientists probably had the broadest set of scrapie strains of anywhere in the world. They had dozens of different strains of scrapie from sheep, goats, and mice. And they had a treasure trove of experimental methods that had been gleaned from decades of top-notch research.

Gajdusek's visit to Compton occurred around the same time that a brilliant series of experiments on transmitting scrapie to mice was beginning to look promising. Richard Chandler, an expert with inbreeding mice to achieve reproducible traits in offspring, was carrying out the work. Using several different lines of inbred mice, he inoculated different mouse lines with the "drowsy" as well as the "scratching" variants of goat scrapie. Although Chandler's success would not be published until 1961, Gajdusek had access to the unpublished data that showed it was possible to transmit scrapie from sheep or goats to mice.

But Chandler was not finished. Within a few months, he showed that he could now transmit the scrapie agent from mouse to mouse and with far shorter incubation times, a few months in some cases. The brains of the infected mice looked exactly like the brains of scrapie-infected sheep or goats. They were riddled with small holes. Finally, Chandler showed that he could take the scrapie agent from mice and reinfect goats with the mouse variant. Chandler's experiments were groundbreaking because they showed the species barrier could be penetrated by the scrapie "virus," and once in mice, the "virus" adapted very well to its new hosts. The mouse-to-mouse scrapie agent killed mice with much greater efficiency than the goat-to-mouse variant. This ability of the agent to become even more lethal when it jumped species would not go unnoticed by the Compton researchers.

While in Britain, Gajdusek became aware of another intriguing set of findings, which were known in Iceland but had not been well publicized outside the veterinarian literature. In the 1950s, Björn Sigurdsson, a virologist and pathologist, had been studying a set of diseases, dubbed "rida," in sheep. In Icelandic, rida simply meant ataxic or immobile. Rida appeared to have sufficient parallels with scrapie that in hindsight they were synonymous.

The prolific Sigurdsson was also studying a second disease of sheep that had different symptoms from rida but which also took months and years to manifest. The problem could be traced back to a group of Karakul lambs that were imported into Iceland in the 1930s in order to improve the wool by breeding with the indigenous Icelandic sheep. After several years, sheep on the recipient farms began to behave strangely and began dying of respiratory distress. About 10 percent of these sheep became demented and wasted. Sigurdsson proposed that a virus called visna was causing the respiratory/dementia symptoms.

The rida agent obviously differed from the visna agent since the latter primarily caused respiratory symptoms with only a small percentage of the animals succumbing to dementia. Sigurdsson postulated that both visna and rida might be caused by a novel group of slow viruses that had different family members, each giving rise to the symptoms of visna and scrapie. Sigurdsson adopted a program of culling the sheep infected with either the visna or rida "viruses." He noticed, as others had, that sheep newly introduced onto a pasture that had housed rida-infected sheep years before eventually succumbed to rida. This transmission from the soil to newly introduced sheep never happened with visna-infected sheep. This, of course, was a classic feature of scrapie. Sigurdsson concluded that both visna and rida were caused by slow viruses, but that the rida virus was far more hardy and long lasting than the visna virus.

Gajdusek decided to stop over in Iceland on his way back to the United States in order to obtain firsthand experience of the different symptoms of rida (scrapie) and visna. Before leaving Compton, however, Gajdusek gathered a group of vials, each containing different strains of scrapie from goats, sheep, and mice, and simply put them in his pocket. He simply disregarded the strict new USDA guidelines for importation of any scrapie into the United States. They amounted to what was, in effect, a strict embargo. Gajdusek ignored these rules and would later enter the United States with a diverse collection of scrapie strains in the pocket of his jacket.

Gadjusek returned from his trips to England and Iceland convinced that the key to kuru lay in isolating a "slow virus." This was the beginning of Gajdusek's long campaign to promote the "slow virus" explanation for kuru, CJD, and scrapie.

No matter how long the incubation time of the deadly kuru agent, it was now clear to Gadjusek that the transmission experiments were the next priority. Smadel reluctantly agreed. Inoculating animals would mean a complex series of logistics, including purchasing primates, finding a suitable facility, and finding a top-notch, can-do administrator and project manager. Smadel knew that Gajdusek was much too valuable to babysit chimpanzees for several years.

At this juncture, it was also obvious that Gajdusek's insistence on conducting field autopsies in the Highlands of New Guinea had paid off. There was now a plentiful supply of kuru-infected brains in the United States that could be used for infectivity experiments. The kuru brains were stored at Bethesda and later at Fort Detrick, which was, and remains, the center of United States biological warfare research.

One of the few descriptions of this impressive collection of infected brains comes from an account by the young medical student Robert Klitzman as he was escorted into the high-security facility at Fort Detrick in the 1970s. Along a wall inside a keypad-secured room in Building 84, ". . . shelves upon shelves supported lidded circular glass tanks the size of ladies' hat boxes. In each jar a human brain floated, infected . . ." he was told, with kuru. "The organs swam in pale green fluid, suspended on their sides or facing forward, sunk to the bottom or pressed against the glass, as if staring out. Scotch taped to each glass drum was a white index card labeled with letters and numbered in red magic marker."

The kuru brains that Klitzman saw at Fort Detrick were to be the main source of material for the infectivity experiments. Klitzman explained the basics of creating the brain slurry: "I mixed masses of kuru infected brains with various chemicals to form a clearish pink extract that had to be stirred for several days. I poured the mixture into a tall Ehrlenmyer [sic] flask—with a round bottom and a tall narrowed neck that impeded fumes from escaping—dropped in an oblong white plastic magnet, and placed the container on an electric plate that made the magnet spin around to stir the soup of brains. I set the whole apparatus in a tall glass-fronted refrigerator where the magnet would turn for several days."

Meanwhile, Gajdusek and Smadel set about finding a facility in which to begin their transmission experiments to primates. Gajdusek even went so far as to say he needed a facility "where nobody could see what we were doing," he later told Pulitzer Prize–winning journalist Richard Rhodes. They finally settled on a couple of remote and abandoned farm buildings and barns (one large and one small) located deep within a couple thousand–acre wildlife research facility located just off I-95 between Baltimore and Washington, D.C. The facility was called Patuxent. The place had several advantages. Its proximity to Washington, Bethesda, and Fort Detrick meant that it was easily accessible to Gajdusek when he was in the country, while the large, densely wooded preserve kept all activity there away from prying eyes.

But Patuxent itself was by no means abandoned; only the barn buildings in a remote section of the facility had fallen into disuse. The Patuxent Wildlife Research Center had been around since 1939, and by 1960 it was a national center for studying wildlife migration patterns. According to a 1959 brochure, "Facilities at Patuxent now include a wildlife pathology laboratory with animal pens and aviary, a biochemical laboratory, a bird distribution center . . . personnel of the station includes about one hundred employees, about forty of whom are professional biologists, chemists or statisticians."

All Smadel and Gajdusek needed now was a scientist to run their inoculation program. Their first choice was obvious—Hadlow. Now considered one the world's experts on scrapie, Hadlow had returned triumphantly from his research sabbatical at Compton and in 1961 had secured a position at the Rocky Mountain Labs in remote western Montana. After one of Smadel's recruiting trips, flying from Washington to Hamilton, Montana, Hadlow fondly recalls Smadel's greeting comments, so revealing of his well-known abrasive personality: "Godammit, Bill, I can get to Rome quicker than I get to Hamilton, Montana." But Hadlow had just moved to Montana, his wife had given birth to a baby, and the family was not about to up and move again. Hadlow refused Smadel's persuasive pitches, though he made it a point to drop in to Smadel's office for coffee anytime he was in Washington, so great was his admiration for the man. Hadlow said he enjoyed listening to Smadel holding court in his office. It was always "Goddam this and goddam that," Hadlow chuckled.

Gajdusek and Smadel eventually settled on Clarence Joe Gibbs Jr., a veterinary pathologist and one of Smadel's protégés, for their can-do administration of the kuru transmission program. Gibbs quickly agreed, under pressure from Smadel, to cancel his sabbatical to Brazil to move to Patuxent, though the location for the most intensive series of kuru inoculation experiments in history still had literally no facil-

ities. So between 1961 and 1963, while Patuxent was being secured and the new buildings were being upgraded, Gibbs went to work at NIH facilities, inoculating hundreds of mice in an attempt to replicate the successful scrapie transmission experiments at Compton. Gibbs also was attempting to transmit kuru to mice. Of course, Gibbs was using many of the scrapie strains that Gajdusek had so casually brought over from Compton in his pocket.

It fell to Gibbs and Gajdusek to inject the first chimpanzees with kuru, a task neither of them relished. Gibbs had purchased three two-year-old chimps. The first experiment was timed for one of Gajdusek's transient visits to the United States. Because the facilities at Patuxent were still not ready, the experiment occurred at the NIH. On February 17, 1963, Gibbs and Gajdusek drilled a small hole into the skull of a chimpanzee they were later to name Daisy. Less than 1 cc of the brain slurry from a newly thawed kuru brain was sucked into a small syringe and injected straight into the anesthetized chimp's brain. Daisy did not appear to suffer any ill effects from the treatment. In the months that followed, they inoculated two more young chimps.

By summer 1963, Gibbs had transferred the vast majority of the small animals, mostly mice infected with scrapie, to Patuxent and housed them all in the upgraded facilities at the wildlife refuge. The kuru-infected chimps followed and were housed in a new cinder block "primate facility" that held over seventy other primates, including African green monkeys, squirrel monkeys, and rhesus monkeys.

Even though the primates were the really important test of Bill Hadlow's suggestion, it was necessary in the years that followed to conduct massive parallel experiments because of the long waiting period between inoculation and the manifestation of symptoms. So in addition to the chimp experiments, a battery of inoculations was carried out at Patuxent and at NIH on gibbons, capuchins, spider monkeys, squirrel monkeys, marmosets, woolly monkeys, baboons, bonnets, bush babies, patas, pig-tailed macaques, sooty mangabeys, and, in the nonprimate category, ferrets, minks, gerbils, goats, sheep, voles, rats, mice, skunks, and guinea pigs. Dogs, cats, chickens, ducks, and geese were also inoculated with kuru, CJD, and scrapie in a wild attempt to find a successful transmission. In short, a huge burst of experimentation was conducted to determine the limits of the infectiousness of this dread new disease. The Patuxent facilities housed many of the animals used in these experiments, as well as numerous other free-roaming animals that were part of the natural wildlife refuge.

That this was hardly the ideal place for testing the inoculations of a potentially dangerous infectious agent did not go unnoticed. Dr. Robert C. Reisinger, who was assistant to the director of the Animal Disease Eradication Division at the USDA, visited Patuxent in 1963 in response to a request from Gibbs for a USDA importation license for additional scrapie strains from Compton. When Reisinger arrived at Patuxent to inspect the containment facilities, according to journalist Richard Rhodes, he was astounded. An examination of his extensive publication record shows that Reisinger was a skilled veterinary researcher who had spent years studying infectious disease in animals. He had published widely on infectious diseases including rinderpest, parainfluenza virus, and myxoviruses in cattle and Marek's disease virus in chickens. His previous work at the USDA with rinderpest, also known as cattle plague, a highly contagious viral disease of cattle, had given him a lot of experience working with infectious organisms. So contagious was rinderpest, in fact, that the transport of infected animals was limited to prevent widespread outbreaks of cattle plague. In short, Reisinger was thoroughly knowledgeable about the necessity for containment of infectious disease when conducting research.

Reisinger was also aware of the embargo that the USDA had placed on the importation of scrapie into the United States, according to 1962 and 1963 memos found at the National Archives. These memos create a picture of Reisinger as a conscientious official who in 1962 and 1963 dominated the import and export of live biological samples into the United States and was involved with the USDA's scrapie eradication program. Since the 1950s scrapie had become an increasing problem and had led to widespread slaughter of infected flocks. It was quickly apparent to Reisinger that Gibbs and his group were working with illegally imported strains of scrapie at Patuxent and most likely at NIH prior to that. Not only were they defying common containment rules with an agent that was on high alert from the USDA, they were also injecting these organisms, as well as kuru, into a wide variety of animals. Reisinger would have been well aware that kuru had already killed hundreds of humans in New Guinea and that here kuru was being injected into primates, monkeys, and dozens of other animal species.

Even worse, Reisinger couldn't have missed just how crowded the shared lab facilities were at Patuxent. One researcher from those days at Patuxent told me that the kuru group shared lab facilities with several wildlife research teams, including a team of researchers from Walter Reed Army Hospital and possibly a group from Johns Hopkins University. It seems unimaginable that the transmission of kuru was occurring under the same lab roof as other wildlife experiments, but at the time there was skepticism that the efforts to spread kuru by inoculation would succeed, but at the same time there was skepticism that the efforts to spread kuru by inoculation into different species would succeed.

Reisinger viewed the safeguards at Patuxent as so substandard that he considered it a top priority to shut down the facility as soon as possible. According to Reisinger, containment was totally inadequate at Patuxent. Gibbs, for his part, claimed that because the diseases he was dealing with were not contagious, Reisinger's concerns about mice escaping from the facility were unfounded. Later research on the transmissibility of kurulike diseases would call his conclusions into question, of course, but at the time he was convinced that there was no danger of such diseases escaping into the outside world.

Within a short time, Rhodes reports, Reisinger had written a letter to the director of the NIH reporting the deficiencies at Patuxent. He specifically cited Gajdusek's illegal procurement and transport of multiple exotic scrapie strains from Compton into the United States. He advised that the facility at Patuxent be shut down forthwith.

Unfortunately for Reisinger, the committee he put together to review the biosafety aspects of Patuxent was composed of good friends and supporters of Gibbs. Smadel was not one of them, however, having succumbed to cancer on July 21, 1963. In an effort to appease Reisinger, Gibbs moved his mice to the basement of Patuxent, where they could be more fully contained. Reisinger, however, pushed forward with his efforts to close Patuxent. In response, Gibbs says that he "sort of arranged to have him offered a position at the National Cancer Institute and off he went."

But Reisinger's exit could not hide the truth: The Patuxent wildlife facility, at least at the outset, was a study in inadequate containment. Although accounts from the facility fail to provide any smoking guns, one must wonder how many inoculated simians defecated unknown titers of kuru agent onto the ground when they were let out to play or escaped. How many deer subsequently grazed on the grass once occupied by infected animals? Did mice or other inoculated animals, as Reisinger feared, escape into the wildlife refuge at Patuxent?

Even after Smadel's death, the work continued feverishly at Patuxent and at the NIH facilities both at Bethesda and at Fort Detrick. Joe Gibbs had expanded his team significantly and by 1965 he had created a large network of interested researchers. The number of staff at Patuxent rose to about twenty, including animal handlers.

Gibbs had recruited an excellent team to monitor the animals at Patuxent, Mike Sulima among them. Sulima had been Gibbs's senior technician for many years. Sulima would spend hours sitting looking at the chimps playing and often took them to the area outside. He knew their individual personalities as one might know the personalities of a group of children. He grew very fond of the chimps through his daily interaction with them. The chimp colony quickly grew to more than twenty members. This was not counting the numerous other members of the menagerie that was rapidly becoming a full house.

Sulima remembers that Gibbs did not really expect the kuru transmission experiments to be successful. "Even a negative experiment is valuable, he would tell us all the time," Sulima told me in March 2004. From Gajdusek's correspondence with Gibbs, it was also plain that the chief did not expect kuru to be transmissible either.

Two years was all it took to prove them both wrong. Sulima was the first to spot a change in Georgette, the third chimp to be inoculated in September 1963. "Georgette was always very friendly," Sulima told me, "she loved to play, and then one day I noticed she began to withdraw. She started sitting by herself with a strange expression on her face." This was sometime in late June 1965, Sulima remembered. As time went on, Georgette's symptoms became more pronounced. When she was allowed out in the play area she began to stumble slightly, and every so often she would tremble as if a cold breeze had hit her.

Sulima notified Gibbs, who was returning from a conference, that something did not seem right with Georgette. Gibbs immediately began to study the animal. Georgette's normally playful personality had vanished, her lower lip had begun to droop. By the time Gibbs decided to cable Gajdusek in remote New Guinea that Georgette appeared to be showing signs of kuru, Daisy, the first chimp to be inoculated, had also begun to change.

Daisy began exhibiting the same pattern of unusual behavior they had seen in Georgette. "Their behavior was the first to change," Sulima noted. "The aggressive ones would become quiet and the quiet ones would start to become aggressive." These behavioral changes would be followed by tremors, unsteadiness, and a loss of balance as the kuru increased its deadly grip on the animal's brain.

Gajdusek flew the red-eye back from New Guinea and immediately drove to Patuxent to see for himself what was happening. He had seen kuru patients in the New Guinea Highlands less than seventy-two hours previously. Now in the United States, he couldn't believe his eyes. Bill Hadlow's prophetic *Lancet* paper was staring at him through the vacant eyes of a couple of chimps. The kuru universe had suddenly expanded enormously.

6

Mad Mink

E ven though Bill Hadlow had refused the job offer to oversee the kuru inoculation experiments at Patuxent, he did not remain idle. In the summer of 1963 a veterinarian from Idaho reported that mink appeared to be showing signs of neurological disease and for demonstration purposes an animal was brought to Hadlow for inspection at the Rocky Mountain Labs in Hamilton, Montana. Hadlow killed the animal and conducted a necropsy. "All the hallmark changes were there," he reports—the holes in the brain cells, the degradation of the neurons, and the lack of inflammation. Hadlow's interest was piqued. The mink showed the classic patterns of scrapie that he had seen under the microscope countless times before. He would later call the disease transmissible mink encephalopathy, or TME, but in 1963, nothing was known about this disease.

Quickly, Hadlow delved into the historical veterinary

records looking for reports of mink in the United States showing neurological disease and soon realized that he had hit yet another gold mine. "It [TME] was identified as a new disease of ranch mink in 1947 in Wisconsin by Gaylord Hartsough, a veterinarian with an uncommon understanding of the mink industry," wrote Hadlow nearly four decades later in a commentary that he told me was his swan song. "At that time, the characteristic degenerative changes seen in the brains of affected mink were regarded as those of a neurotoxicosis, presumably the result of exposure to an unknown noxious substance on the farm. When the disease appeared again in 1961 on five farms in Wisconsin sharing the same food supply, the feed became the most likely source of the noxious substance responsible for disease on each farm." In 1963 two more outbreaks of TME occurred in Wisconsin, and others were reported throughout the United States.

Prompted by his findings that TME appeared to have had a long history in the United States, albeit unrecognized, and that the neurological signs in the animals were unmistakably those of scrapie, Hadlow decided to begin inoculation experiments with mink in Hamilton, Montana. By late 1964 and early 1965 Hadlow had shown that TME was transmissible to mink by injecting the brain slurry of one mink into the brain of a second animal. Within a couple of years he had also shown that TME was transmissible to several other animal species. Once again Hadlow had demonstrated the originality that is central to groundbreaking research. Not only had he first announced that scrapie and kuru were similar diseases, he had also suggested the pivotal transmissibility experiment—the inoculation of chimps with slurry from kuru brains. And now a few years later, Hadlow had opened up yet another front in the discovery process by showing that TME was an example of the ever-burgeoning family of "slow virus" diseases.

By this time, a young veterinarian by the name of Richard Marsh was on the case. Marsh was the son of a Wisconsin mink farmer. He had an intimate knowledge of the behavior, feeding habits, and illnesses of these unusual animals. Like most mink farms, the Marsh farm had been a family business for several generations.

Mink have been farmed for fur in the United States since shortly after the Civil War. Throughout their history, mink farmers have employed selective breeding to develop a wide variety of pelt colors, many of which are unknown in nature. For decades, the state of Wisconsin had led the country in the number of mink farms and production of the much sought-after mink pelts. Wisconsin also happens to be a major center of the dairy cow industry.

During the mid-twentieth century, following the gradual introduction of more efficient farming methods, a change took place in the mink farming industry. The increased drive toward efficiency meant a greater focus on more economic feeding practices. In the wild, mink are voracious, carnivorous hunters. They regularly eat fish, birds, frogs, and small rodents. In captivity mink were fed different diets, including leftovers from fish-processing plants and some rendered animals. In Wisconsin, as the economics of feeding mink began to assume greater importance, many mink ranchers began to scour the adjoining countryside for the remains of dead cows to grind up and feed to the mink.

Following in the tradition of mink-farm children, Marsh studied at Washington State University and became a veterinarian in 1963. Quiet and unassuming, Marsh continued his studies and began to work as a graduate student at the University of Wisconsin–Madison. Madison was a natural locus for a research drive into this obscure disease, since, as Hadlow had discovered, the concentration of mink farming in Wisconsin had already given rise to a couple of TME epidemics in the state. And given such an obscure field of research, it was natural that Marsh and Hadlow would meet, and Marsh was quick to begin to follow up on Hadlow's transmissibility experiments with TME. By the end of the decade Marsh and his professor, Robert Hanson, had shown that TME was transmissible to squirrel monkeys and that, more important, scrapie was transmissible to mink. The finding had some implications for the entire burgeoning industry of mink farming because it was not unknown for mink to be fed the ground-up remains of dead sheep.

After the successful transmission of scrapie to mink, Marsh began an exhaustive investigation of previous outbreaks in the United States and Canada. Marsh contacted two veterinarian researchers who were expert in mink encephalopathy, Gaylord Hartsough and Dieter Burger. Hartsough, the veterinarian who examined the 1947 outbreak, and his colleague Dieter Burger had carefully investigated the feeding patterns in a TME outbreak that had occurred simultaneously at two adjoining farms in Canada in 1963. The owners in Canada were adamant that no dead sheep had gone into the feed. Rather, in both cases they had fed the mink ground-up remains of the same downer cows.

Downer cows are simply cows that were too sick to stand or walk and usually they are put out of their misery. If the meat from downer cows was judged unfit for human consumption, it was usually ground up and fed to animals. According to the Canadian ranchers, the only thing their mink had been fed was the ground-up remains of downer cows. Marsh scratched his head. At that time no scrapielike disease was known to exist in cows.

Going back to the original 1947 outbreak and following each subsequent sporadic outbreak, Burger and Hartsough noticed that every disease eruption appeared to have something in common. Each occurrence was usually manifested on a couple of adjoining ranches that shared common feed. But in the 1963 outbreak it was clear that rendered cows, and not sheep, had been the dominant source of feed for the mink. Certainly the evidence looked very suspicious. Hartsough and Burger began to suspect something worrying. They did not take long to verbalize it.

In December 1964, Gajdusek organized a conference at NIH that focused on what was known about kuru, scrapie, CJD, and related diseases. For the first time the data from the early mink transmissibility experiments were presented at the conference and Gajdusek was quick to see that the family of hosts for these mysterious diseases had expanded once again, into mink. According to the conference proceedings, Hartsough was blunt in his assessment of TME's origin: "It appears that these mink were fed beef, and it is conceivable that the disease is caused by a virus which is commonly present in cattle."

Here was Hartsough, a solid professional at a gathering of

the premier kuru/scrapie/CJD researchers in the world, strongly suggesting that the Canadian mink outbreak was the result of a new scrapielike disease in cows. His statement was speculative, of course, based only on circumstantial evidence that the diseased mink had been exclusively fed with ground-up cows, but it foreshadowed the grim reality that would unfold in the next couple of decades. Marsh, for his part, would doggedly pursue this line of research—and end up risking his career for it.

7

Cannibalism

Meanwhile, Gajdusek continued his commute halfway around the world. He was more than content to allow Gibbs to hold the fort at Patuxent and to return occasionally to the United States to review the data on the inoculations. This arrangement left him with ample time to indulge his passion for life in New Guinea. Every year he would spend several months living among the Fore and the other tribes that he loved.

Gajdusek's travels through the villages became the Fore equivalent of major media events. Whenever Gajdusek hiked through the Highlands, word would spread from village to village that the white man who "travels great distances" was in the area. Whole villages would turn out to greet him as he passed through. His larger-than-life personality quickly attracted a retinue of boys who carried his bags and ran ahead of him, clearing difficult passages and even constructing makeshift bridges across dangerous gorges. Some boys who were especially skillful would help Gajdusek treat the kuru victims in the villages, and some even helped him perform makeshift autopsies far away from any modern medical conveniences.

Gajdusek, whose career began as a pediatrician, treated many of the children he encountered in New Guinea for a wide variety of ailments, such as parasite infections and infectious diseases besides kuru, which were prevalent and amenable to the interventions of Western medicine. And because many children were orphaned by the devastation wrought by kuru, in 1963 Gajdusek also began to adopt some of the children he befriended. His intent was to repay some of the kindness shown to him by the Fore by bringing them back to the United States and paying for their upbringing and education.

Overall, Gajdusek would be credited with sending more than thirty children through the American educational system. They lived either in Gajdusek's own boyhood home in Yonkers, New York, or in his three-story colonial, complete with fruit trees and an Olympic-size swimming pool, set on a 100-acre knoll overlooking Frederick, Maryland. Many of his highly educated, adopted boys would later return to New Guinea and Micronesia, where they made substantial contributions to the welfare of their homeland.

With his deep and abiding love of ethnography and anthropology, Gajdusek was profoundly concerned with the double-edged sword of Western intervention into the lives of these "primitive" people of New Guinea. While the medicines actually did improve the lives of people, whose infant mortality rate was astonishingly high, it was impossible not to disturb the idyllic culture of the New Guinea people. "We cannot cure their yaws and ulcers," wrote Gajdusek in his journals, "save their dying children, remove their arrows and treat their wounds without coming to them. We cannot come to them without bringing ourselves and our life into their horizon and to then refuse their request to see the outer worlds, or agree with those who would come and study them, observe them, and especially those who want to 'help' or change them, in any way (including to stop warfare, murder, fear, superstition, famine, or pestilence) and who would yet 'leave them as they were primitive and picturesque' . . . is an insult to their human aspirations and intelligence and will never do. By coming we commit ourselves to the change and are agents of it. The change disturbs us for we know better than they do how pallid and barren and how unsatisfying the fruits of civilization can be at times."

Because of his high regard for the Fore people and the love that he felt for their customs, Gajdusek was fiercely protective of them. He strongly defended them against the tendency of the Australian press to write off the Fore as "Stone Age savages," and he defended against the mockery of their lifestyle and customs that sometimes prevailed in the Western press. He also dismissed the idea of a connection between cannibalism among the Fore and kuru, though the possibility of such a link did occur to him early on. "A number of wild possibilities come to mind—some can be dismissed already," he wrote in an intriguing missive to Smadel on November 24, 1957. One possibility he mentioned involved cannibalism—that eating brain may "sensitize" the system and make a person more susceptible to illness later on. Wrote Gajdusek: "The rub: cannibalism has ceased, did not involve brain as far as we can find, and even if rare cases still occur, many of our youngest patients have rather certainly not consumed human tissue. . . ."

In the early 1960s, Gajdusek thought that kuru might have a hereditary component. The disease seemed to be very closely associated with mothers and their children. It was much less obviously associated with the male members of the tribe. A popular theory at the time was that kuru was a hereditary disorder with a gene that was dominant in females and recessive, or less prone to be expressed, in males. But if that were the case, it would be difficult to imagine a gene spreading that rapidly through the Fore population unless it conferred some extreme survival advantage.

The possibility of a hereditary link to kuru would be shattered by one of the many teams of anthropologists who arrived in New Guinea in the wake of Gajdusek's larger-than-life reputation and his world-renowned kuru research. The husband-and-wife team of anthropologists Robert Glasse and Shirley Lindenbaum were funded by the Rockefeller Foundation to pursue genealogical research among the Fore. Lindenbaum and Glasse arrived at a village called Wanitabe in 1961, quite fortuitously, as it turns out. There the local tribal elders embraced the couple with open arms because they fulfilled a cargo-cult prophecy that predicted the arrival of white people bearing new gifts.

As Glasse and Lindenbaum progressed in their fieldwork, the Fore began to accept them, and the more the bonds of trust were fostered, the more the Fore began to confide in

them. The process eventually led to an understanding of the
Fore's concept of time, which was extremely vague by Western standards. This fuzziness meant that no one had an accurate history of the illness. The Fore tended to focus only on
the few days around the present, with only cursory attention
to events in the past. To get around this problem, Lindenbaum and Glasse probed the origin of kuru with questions
related to the age of individual tribe members, triangulating
in time the approximate time of birth, of initiations, of marriages, and then cross-referencing these memories with events
that happened that had an independent date.

One such event was the crash of a Japanese airplane during
World War II. Glasse and Lindenbaum tried to obtain the
dates of marriages, initiations, births, and deaths relative to
that crash and gradually they began to suspect that kuru was
a relatively recent disease. Some of the older Fore appeared to
remember when kuru first appeared, maybe in the 1920s or
1930s. The disease then accelerated in the 1940s and spread
west and south through the Fore region. When Lindenbaum
and Glasse began their fieldwork in 1961, so great was the
death toll from kuru that some hamlets had a three-to-one
ratio of males to females and there were large numbers of
orphaned children.

If the kuru was actually that recent, only a few decades
old, Glasse and Lindenbaum reasoned, then it was unlikely to
be hereditary because it had spread much too fast through the
population. If kuru had spread rapidly through the Fore population, which comprised about 40,000 individuals, then the
idea that all Fore people were descended from a single individual only a few decades previously made no sense. The

main caveat to this argument, of course, was the reliability of the older Fore memories, as well as their nonlinear concept of the arrow of time. As anthropologists, Lindenbaum and Glasse were very cognizant of a well-documented tendency in many groups and tribes to mix and match legend and myth with fact when talking about anything in the past going back more than a few years. Nevertheless, as Glasse and Lindenbaum probed further, and as they cross-indexed scores of narratives, they obtained more information that appeared to substantiate a relatively recent emergence of kuru.

Before long, Robert Glasse and Shirley Lindenbaum's patient research also began to reveal some surprising information about Fore cannibalism. Shirley Lindenbaum slowly gained the trust of the Fore women. Unlike the local missionary's wife and daughter, who never ventured beyond the huge fence that enclosed the mission, this white woman even let the Fore women feel her body to reassure them she had body parts just like theirs. And within a few months of keeping the company of Fore women, Lindenbaum began to gain access to secrets and taboos that Gajdusek had never been able to penetrate.

Little by little, Lindenbaum began to extract some remarkable information from the Fore women. She discovered first of all that cannibalism also was a relatively recent Fore custom. From interviews with dozens of Fore women, Lindenbaum's estimate was that cannibalism had probably begun sometime in the early twentieth century. The second surprise, no doubt, was that it was primarily practiced by women.

Lindenbaum learned that Fore society was one of huge

gender disparities. Fore women lived in a different world from their men, and this different world had some very secret customs. In a remarkable discussion with Richard Rhodes, author of *Deadly Feasts*, Lindenbaum revealed the shocking details about the secret cannibalism practiced by Fore women: "They hid in the garden, they did it at night, did it away from the eyes of people, they did it in the old sugarcane gardens and sometimes even in the burial ground. It stirred erotic excitement and maybe even gender power. Women were attacking men's bodies. It was their own domain and men were kept out."

Gradually, it emerged that the cannibalism occurred with children in attendance and the women made a practice of scooping the brains out of the skull and eating them raw. The children were allowed to try these delicacies and brain tissue was routinely smeared around their mouths and on their hands. Without standard Western hygiene procedures, the presence of smeared brain tissue remained on their hands and faces even several days after a cannibal feast had taken place. And when the men did participate, Lindenbaum learned, their exalted status in the tribe allowed them the choicest morsels, comprised primarily of muscle. Hence the women and the children were devouring the nervous tissue while the men occasionally ate muscle.

Another talented scientist to arrive on the scene early, again in response to Gajdusek's rapidly escalating fame and notoriety, was Michael Alpers, a young Australian physician. In late 1961, Alpers was dispatched on a two-year Australian government contract, probably as a hasty political countermeasure against what they perceived as the loud-mouthed American who had barged into Australian territory and commandeered their project. McFarlane Burnet and his team at the Walter and Eliza Hall Institute in Melbourne were conscious of having ignored the pleas from Vincent Zigas for help back in 1955. Alpers was thought to be the Australian solution to the "problem" of Gajdusek dominating kuru research. Alpers was initially placed in the middle of the kuru project in order to be the Australian eyes and ears. His duty was to wrest some of the control of the kuru research project back into Australian hands.

Alpers was considerably quieter and more introspective than his American counterpart, and he and Gajdusek soon established a friendly working rapport, quite contrary to all expectations and intent. Zigas describes Alpers's arrival on the scene: "The last entrant to join the agglomeration of new faces was a young doctor from Adelaide, Michael Alpers and his wife and baby daughter. Unlike Carleton, Alpers was a taciturn type with a philosophy of listen, look and be silent. He was a supremely confident yet self-effacing man, gracious in manner, polite in speech, but implacably stubborn. . . . Despite the difference in temperaments, Michael and Carleton fit each other well. Carleton had absolute confidence and trust in Michael."

Alpers soon began to work seriously on the epidemiology of kuru, drawing from Gajdusek's extensive experience in the field. By the time Alpers joined the medical investigation, Gajdusek's reputation for tireless epidemiology was unrivaled. "This was an extraordinary epidemiological survey, of a kind that only a person like Gajdusek could pull off, because of his mad energy and his mad adventurousness," remembers Lindenbaum of one of Gajdusek's field trips, ". . . [his willingness] to disregard the necessity to get permission from the authorities to do certain things, as well as go in and out of areas . . . he was going to be a headache for them in any case. He doesn't get permission for any of these journeys, of course, nor for any of the medical procedures—he just goes off and does them."

Together Gajdusek and Alpers crisscrossed the kuru areas of the Fore territory, painstakingly tracking down family trees, locating family members who had moved or had married elsewhere, and gradually assembling a thorough and complete picture of hundreds of Fore genealogies, often through several generations. Theirs was the most thorough and detailed epidemiology ever conducted on kuru. Soon it became obvious, even in the field, that kuru appeared to be less severe in younger children than it once had been. There were tantalizing clues that the disease might have been receding gradually in the 1960s, but nothing that could be called statistically significant. In areas where European culture had penetrated more aggressively, for instance, there appeared to be fewer children dying of kuru, and in areas where the Australian program of "pacification" had been most successful, kuru seemed to be less of a problem.

After his two-year apprenticeship with Gajdusek, Alpers moved to the United States and became an integral part of the operations at Patuxent and at NIH. But just prior to embarking for the United States, Alpers had lunch with Glasse and Lindenbaum. In the time since Lindenbaum had been allowed into the inner sanctum of the Fore women's secrets about the widespread cannibalism, Lindenbaum and Glasse had come across an arresting article in *Time* magazine. The article, which described research suggesting that a planarian flatworm's characteristics could be transmitted by one worm eating another, was later to be proven almost completely wrong. But it is no small irony that an article focusing on pseudoscientific experiments with flatworms was the catalyst that triggered the lightbulb to go on for Glasse and Lindenbaum.

All of a sudden the idea that an infectious agent could be transmitted through the Fore by eating their relatives became a very potent hypothesis. During the lunch, Alpers and Glasse and Lindenbaum engaged in a lengthy discussion about the possibility that cannibalism and kuru might be linked via the transmission of some infectious agent. In fact, Lindenbaum and Glasse had had that suggestion planted in the backs of their minds from the very beginning of their stay in New Guinea, thanks to an intriguing paper by another pair of anthropologists who had since returned to Tulane University in the United States. In 1961 the husband-and-wife team Ann and John Lyle Fischer had published what would turn out to be a remarkably prescient paper on the possible link between cannibalism and kuru. But as usual in the fragmented terrain of science, only anthropologists would have been aware of their report.

"The Fore habit of eating corpses suggests a way in which a viral agent might be passed," the Fischers had written. ". . . Victims of some kinds of sorcery are not eaten by the Fore, who fear they might be poisoned, but kuru victims are evidently not included in this category . . . corpses are said to be consumed in all stages of decay and with all degrees of cooking. If this is the case, then women are probably more likely to eat raw corpse than men are. An infectious agent could explain why kuru seems to be passed down through maternal rather than paternal grandmothers. . . ."

But the Fischers had little to go on—it was really just a guess. Glasse and Lindenbaum were in a similar position. What would finally convince everybody was the demonstration of the infectivity of kuru. Without an infectious agent, nobody took the cannibalism-kuru link seriously.

Alpers arrived at Patuxent in early 1964 in the midst of the large-scale inoculations of chimps and monkeys. He was at Patuxent with Mike Sulima and Clarence Joe Gibbs in June 1965 when Georgette the chimp began to show the first signs of strange behavior. The obvious signs shown by Georgette and later Daisy convinced Alpers to reexamine the voluminous epidemiological records of the Fore kuru deaths, which were then located at the NIH.

Slowly and painstakingly, Alpers went through the thousands of pages of journal notes. He assembled the familial interrelationships and finally he tracked the deaths by age, month by month, of the Fore from the early 1950s all the way through the mid-1960s. The pattern that emerged was unmistakable. Kuru was definitely receding in the Fore and the trend in decreasing mortality was most dramatic in children. Now that the first few chimps had come down with kuru, the last piece of the puzzle, the infectiousness of kuru, seemed to be in place. As the 1960s wore on, the epidemiological evidence seemed to be building that kuru and cannibalism were linked. More of the specific feasts of the Fore were now being placed in time by more in-depth interviews, and knowledge about the cessation of kuru in children had become dramatic by 1967.

Alpers's eureka moment occurred in 1967. "One morning," he told Richard Rhodes,

> working at home in my makeshift office on the porch of our house in Bethesda, preparing a paper for a meeting of the International Academy of Pathology, I was waving all these possibilities around in my mind when suddenly the pieces of the puzzle clicked into place. Cannibalism, though it seemed to be a plausible explanation, did not on close analysis explain the phenomenon as an independent mechanism. However, cannibalism as the single mode of transmission of the transmissible virus of kuru did make sense: it was suddenly all too painfully obvious. It was obvious because nothing further needed to be explained, and painfully so because of the agonies of uncertainty that existed when the explanation seemed so close at hand and yet not quite there: until that moment when it all clicked into place. But what a moment!

Shortly thereafter, Glasse and Lindenbaum also broke through the insular world of anthropology journals by publishing their paper, simply titled "Kuru and Cannibalism," in the widely distributed British medical journal *The Lancet*. Their article explicitly laid out the kuru-cannibalism connection.

Meanwhile, Igor Klatzo's original suggestion linking kuru with CJD was put to the test. Buoyed by the success of the kuru inoculation experiments in chimps, the Patuxent group of Gibbs, Alpers, and others decided to close the loop. The reasoning was that if CJD was the same disease as kuru, and kuru was transmissible, then CJD should be transmissible also. In late 1966 chimpanzees were inoculated with brain slurry from a patient who had died of CJD, and just over a year later the results were in. They were unequivocal. The paper, written by Gibbs, Gajdusek, Alpers, and their colleagues, appeared in the July 26, 1968, issue of *Science* and in relatively simple scientific language described their success:

> . . . inoculation of brain biopsy material from a patient having Creutzfeldt-Jakob disease, with severe status spongiosus into a chimpanzee was followed after 13 months by the appearance of a subacute, progressive, non-inflammatory, degenerative brain disease. The clinical course of the disease was not unlike that in the human patient, and the neuropathological findings were remarkably similar. . . . We believe that Creutzfeldt-Jakob disease has been experimentally transmitted to the chimpanzee, and that the disease is caused by a transmissible agent.

The experiments at Patuxent had proven beyond a doubt that both kuru and CJD were transmissible. Gajdusek was now a highly respected leader of one of the best-known NIH research teams in the world. The group had published hundreds of papers on kuru and CJD. And Gajdusek must have finally come to accept the incontrovertible mountain of evidence supporting the hypothesis that the spread of kuru was by the cannibalism of his beloved Fore. Regardless, it is now generally accepted that the resounding success of the CJD-inoculation experiments sealed the inevitability of a Nobel Prize for Gajdusek.

Gajdusek received the news that he had won the prize in October 1976, while in his family home in Yonkers surrounded by his collection of adopted children. The news was not entirely unexpected. After all, the prize was the reward as much for the unconventional way he had gone about medical research as for his brilliant contributions to it. But it is also obvious that the prize could have been shared with Bill Hadlow for his first linking scrapie to kuru and for setting up the concept of the inoculations at Patuxent, or with Joe Gibbs, who had sacrificed the remainder of his career as a virologist when he was pressed by Smadel into accepting the position of running the inoculation experiments at Patuxent.

In December 1976, Gajdusek and eight of his retinue of New Guinea boys journeyed to Stockholm to collect the prize. The stately traditions of the Nobel Foundation dictated that the Nobel Prize acceptance speech would be forty-five minutes long. As usual, Gajdusek ignored convention and spoke for more than two hours. His wild, rapid-fire speech combined with the visually arresting spectacle of his large retinue of New Guinea boys dressed in formal attire made an unforgettable memory for the Nobel attendees. Gajdusek promised to use the award money to send the boys to college. One of the eight present at the Nobel ceremony was the first boy Gajdusek had brought over to the United States, now named Ivan Mbaginta'o, who was already beginning to establish a scholarly reputation for his research and scientific reports.

The success of the transmission experiments, the linkage of kuru with cannibalism, and the euphoria surrounding the Nobel Prize, however, did not disguise a simple unanswered question that was beginning to resound loudly through the research community: What exactly was the infectious agent? A decade of work by a handful of researchers in Great Britain would soon lead to an answer.

8

Slow Virus

I t was perhaps inevitable that the traditional Scottish-English enmity that erupts every year on the rugby field would carry over into the small field of scrapie research as a rivalry between the Compton and Moredun research institutes. Compton is in Berkshire, England, and Moredun is in Edinburgh, Scotland. Both facilities had enormous prestige in their respective countries. The clashes between scrapie researchers with different data became legendary. The feuding broke into the open at a celebrated symposium in Washington, D.C. in 1964 that was later dubbed the Battle of Washington. According to one attendee at the conference: "The American sponsors were astonished to witness the violent arguments between British scrapie workers, which included dramatic walkouts and scathing criticism of each other's work."

A cultural difference in veterinary science was apparent,

especially during the second half of the twentieth century, between rival scientists in British research institutions versus their more polite colleagues in other countries. I personally witnessed a loud, extended shouting match two decades later at an international conference on veterinary science in Edinburgh between prominent researchers from the Moredun and Rowett research institutes. Two British researchers stood up in the large audience and loudly berated each other's professional work. The red-faced, finger-jabbing exchange of excessive vitriol and insults continued for over ten minutes and was eventually broken up by the conference chairperson. During this barrage of verbal abuse, other international conferees gazed openmouthed at the display of violent hostility. Later, a British scientist assured me that this was business as usual in British agriculture research. Thus the Battle of Washington was not an isolated incident.

No stranger to controversy, Alan Dickinson at Moredun was in the thick of the scrapie arguments at the Battle of Washington. Dickinson had begun extending the work pioneered by Richard Chandler at Compton. Dickinson injected thousands of mice with different "strains" of the scrapie agent, and soon the remarkable reproducibility of the "strains" became apparent. The mice died within the predictable time frame that was characteristic of each strain. They died with very similar brain lesions, although the lesions differed between strains. The same strain held true generation after generation in mice. Throughout the 1960s and 1970s, Dickinson and his colleagues eventually identified more than twenty scrapie strains. So reproducible were the small differences between strains that the researchers could predict which one had been used just by looking down the microscope at the ravaged brain cells of the mice.

But Dickinson went further. He injected a group of mice with a strain of scrapie called 22C. The 22C strain always took a long time to incubate and show symptoms. Then, using another scrapie strain, 22A, that acted very quickly, he injected a small number of the 22C-infected mice with 22A. The rest of the 22C-infected mice were left to develop the disease. What Dickinson found was remarkable. The group of mice injected first with 22C followed by 22A came down quickly with the symptoms of scrapie. The group injected with only 22C predictably took months to develop brain lesions. It appeared that strain 22A competed with 22C in the mice and that 22A was able to overcome the 22C strain, since those mice with the double dose all developed symptoms quickly. Dickinson's brilliant experiment demonstrated that the scrapie strains competed with each other in the same host animal as if they were strains of a virus or bacterium.

Except that scrapie acted nothing like a virus or a bacterium, as Dr. Tikvah Alper would show in the most dramatic fashion in 1967. At Hammersmith Hospital in London, Alper had established a formidable reputation in the expanding field of radiobiology. Radiobiology is the study of the response of biological systems to radiation. The field had grown rapidly after the horrors unleashed at Nagasaki and Hiroshima in the aftermath of the war. The high-energy projectiles of electrons and atomic nuclei released in the chain reaction of the atomic bomb had shown their devastating impact on the living cell in gruesome detail. The stories of thousands of people dying from the effects of radiation in

Japan had convinced the public that the unseen energy from nuclear reactions was lethal.

Since 1950, Tikvah Alper had accumulated an impressive résumé mapping out the precise dose responses of living organisms and their survival in radiation experiments. Her experiments had encompassed bacterial and human cellular survival rates in response to radiation. Alper knew the mechanics of destroying the delicate DNA strands of organisms with radiation.

Alper showed that if you used ultraviolet light at short wavelength, or gamma radiation, you could so damage the DNA of a cell, a bacterium, or a virus that it could no longer replicate. It just died. So when Alper approached the scrapie agent with her bag of tools, she was bringing almost two decades of precision experience in killing viruses and cells with ultraviolet and gamma radiation.

What Alper found was astonishing. In a remarkable article published in the journal *Nature* in May 1967, Alper reported that the scrapie agent was too small to be a virus and, secondly, that it had defied her best attempts to kill it with radiation. In a chillingly prescient statement at the end of her paper, she foreshadowed what would become commonplace several decades later: "If a group of diseases with a similar type of transmitting agent [to scrapie], capable of replication, were to exist, it is clear that standard methods of sterilization by ionizing or ultraviolet irradiation would be quite inadequate to deal with contamination by such agents."

Alper's dry scientific discourse masked a jaw-dropping conclusion. In addition to surviving formaldehyde treatment and in addition to surviving in a pasture for years, the scrapie

agent was almost impervious to lethal doses of either ultraviolet or gamma radiation. To have an indestructible killer
capable of wreaking that kind of damage to human brains, as
the thousands of Fore people had so horribly demonstrated,
was bad enough. But now Alper was proposing the almost
unbelievable possibility that the scrapie agent could replicate
without DNA.

Not only was Alper's proposal in violation of the central
dogma of biology that every living thing replicated by means
of nucleic acids (DNA and RNA), but her paper appeared to
describe a totally unique and novel organism. If it were not
for Alper's long history of extremely careful and replicated
experiments in the field of radiation biology, she would have
been laughed off the stage. But she was an acknowledged
maestro of radiobiology and her paper sent shock waves
through the scientific community.

Superficially, it seemed that Tikvah Alper's work showing
that the scrapie agent appeared to replicate without nucleic
acid might conflict with Dickinson's research showing different strains, thus implying a strain of bacterium or virus. This
apparent conflict was not real, however. An incredibly hardy
and exceptionally tiny virus that was wrapped in a protective
protein coat could explain Alper's conclusion. It was not
unheard of for different viruses to have very different resistance against UV radiation. Or perhaps the scrapie virus was
able to repair its damaged DNA in a way that was much
more efficient than other viruses. If so, then it could recover
from the "lethal" doses of UV that Alper had thrown at it.

Then the third of a trio of surprising scrapie findings
emerged in 1967. Researchers Gordon Hunter and colleagues

from the Institute for Research on Animal Diseases at Compton stumbled on a curious feature of scrapie vulnerability. While trying to isolate the agent from tissue, they used a series of enzymes that digest protein to try to release the "virus" so that it could be purified. They found that the mysterious scrapie agent was much more vulnerable to enzymes that break down protein than it was to UV radiation. In fact, the treatment of the scrapie agent with protein-digesting agents destroyed the infectiousness of scrapie. These experiments strongly implied that protein was central to scrapie's infectiousness.

It was left to an outsider—and surprisingly, a mathematician—to come up with a theoretical framework to explain this rash of unusual findings. Again, the year was 1967. J. S. Griffith was a mathematician at Bedford College, London, who had a strong interest in the brain and particularly in the mechanism of memory. In fact, Griffith had already published several papers on memory and neural nets. It was not surprising, given his intense interest in the biology of memory, that he would turn his attention to Alper's and Dickinson's work on scrapie. After all, the human forms of scrapie, CJD, and kuru, attacked and destroyed memory. Griffith began to ask the question: Is a self-replicating protein completely out of the question? In his 1967 *Nature* article, Griffith aimed to reassure molecular biologists: "There is no reason to fear that the existence of a protein agent would cause the whole theoretical structure of molecular biology to come tumbling down."

Griffith came up with three scenarios to explain how a protein could appear to replicate without contravening the

sacred dogma of molecular biology. All of them had prece-
dent, so as theoretical suggestions they were not that
unusual. Nevertheless, over the years, a consensus has grown
that Griffith's 1967 paper ushered in the "protein only" theory of scrapie replication. Griffith provided a theoretical
framework for the barrage of apparently conflicting data that
had come out of Alper's and Dickinson's research.

A door had now potentially been opened into the true
nature of the infectious particles that Gajdusek had been calling "slow viruses." The terminology had disguised the fact
that scientists had no idea of what was being transmitted
from human to human in the case of kuru, mink to mink in
the case of TME, or sheep to sheep in the case of scrapie. Now
they were increasingly sure it wasn't a virus. It wasn't a bacterium. Maybe that's why it did not stimulate the immune
system like all known viral and bacterial infections. Why was
it able to survive formaldehyde treatment? Why was it
impervious to the extremely high temperatures that had been
applied to it? How could the scrapie agent survive undetected for years in the soil, only to reappear with deadly efficiency after a new flock of sheep were put out to graze in the
pasture?

Could it be a particularly hardy infectious protein?

9

Prions

Strolling through the rather dingy corridors of the biochemistry and physiology departments at the University of California–San Francisco, one finds it obvious that medical research does not focus on aesthetic surroundings. The corridors are lined with rows of freezers and refrigerators, all humming loudly. The paint is chipping off the walls. The graduate students and postdocs sit in tiny offices that are often windowless, grungy, and old. The furnishings are threadbare. Generally, there is an air of extremely hard work in very unattractive surroundings. But down the corridor and around the corner, all of a sudden the paint is fresh and the corridors are gleaming. This is Stanley Prusiner's empire.

In a field renowned for its competitiveness, nobody is more competitive than Stanley Prusiner. The son of a naval officer, Prusiner was born in Des Moines, Iowa, in May 1942. Prusiner senior served as a communications officer during

World War II on the island of Eniwetok, where the first hydrogen bomb was detonated a decade later. As a child, Prusiner bounced around the Midwest, but finally graduated from the University of Pennsylvania with a degree in chemistry and continued there in medical school. He spent three years at the NIH and in July 1972 he began a residency at the University of California–San Francisco (UCSF).

Only two months later Prusiner's future would suddenly and unexpectedly be mapped out for him. A female patient came into his clinic suffering from progressive memory loss, unsteadiness, and erratic behavior. He quickly learned that she had been diagnosed with, and was dying of, Creutzfeldt-Jakob disease. Prusiner had never heard of it.

Prusiner would recall, fifteen years later, the effect that the patient's disease had on him. "The amazing properties of the presumed causative 'slow virus' captivated my imagination, and I began to think that defining the molecular structure of this elusive agent might be a wonderful research project. The more that I read about CJD and the seemingly related diseases—kuru of the Fore people of New Guinea and scrapie of sheep—the more captivated I became."

But like every researcher before him, Prusiner was horrified at the length of time it took to wait for results in this field. The system of medical research in the United States demands rapid research results, followed by rapid publication. The system, dubbed "publish or perish," was not kind to research projects that took years to come to fruition. For a young, aggressive researcher at the bottom of the career ladder, the prospect of having to wait years for publishable results was simply unacceptable. The more Prusiner read about scrapie research, the more he realized that generation after generation of researchers came and went, and progress seemed awfully slow.

In order to begin the project, Prusiner first moved to the Rocky Mountain Labs in Hamilton, Montana, where he learned the basics of scrapie science from Bill Hadlow, the acknowledged master of the art. Hadlow and Prusiner worked on mice because symptoms appeared in them faster than in sheep. By 1978 the pair had worked their way through 10,000 mice. Armed with the basic techniques and methods, Prusiner returned to UCSF to set up shop. He quickly switched from mice to hamsters, after learning of Richard Marsh's original finding that hamsters came down with the disease even faster than mice.

Prusiner then discovered that if the hamsters were observed carefully for the first signs of disease, an experiment could be cut down to a sixty-day turnaround. Running massive parallel experiments with hamsters, the ambitious Iowan moved through the scrapie research field much quicker than his colleagues had ever done. For an experiment to take sixty days rather than nine to twelve months was an enormous leap forward. Prusiner made it happen.

In 1978, Prusiner performed the same ritual that every other successful researcher in this field had performed. He made the pilgrimage to the remote village in New Guinea where Gajdusek spent much of his time. Gajdusek remembers the unathletic Prusiner being practically hauled by native guides into the high mountain village where he lived. On that trip and on a subsequent trip two years later, Prusiner and Gajdusek did the clinical workup on about fifteen kuru patients and Prusiner saw firsthand how the entire field had started. The two brainstormed late into every night.

Back at UCSF Prusiner received an unpleasant shock. The prestigious Howard Hughes Medical Institute had withdrawn their funding and he was informed that UCSF had refused his promotion to a tenured position on their faculty. Prusiner had been caught in the perilous "publish or perish" web. In the midst of an intense research program, the young biochemist was deflated. Without the funds to support the very expensive infrastructure of thousands of mice and hamsters, Prusiner had no bread and butter. Luckily, RJ Reynolds, the cigarette manufacturer, stepped in to fill the gap and enabled Prusiner to continue his research.

Prusiner's accelerated approach paid dividends quickly and he began homing in on a biochemically pure form of the scrapie agent. He quickly achieved a 5,000-fold enrichment of the infectious agent. A pure agent was the holy grail that would finally force the elusive killer to yield its secrets. Using a wide range of biochemical tricks, Prusiner began to accumulate increasing evidence that the infectious particle was composed of protein. No matter how hard he looked, he could not find nucleic acid. Being unable to find nucleic acid pointed away from a virus, even a "slow virus."

By the late 1970s, Gajdusek had also come to the conclusion that the infectious agent did not contain DNA. Gajdusek and Prusiner were in agreement on that point. But Prusiner had a surprise up his sleeve.

Meanwhile, a young graduate student biologist by the name of Patricia Merz was approaching the scrapie problem from a completely different direction. Being a practical young woman and knowing that nobody had ever seen the scrapie agent before, Merz asked why not actually look for it? She chose the instrument with the highest potential for actually seeing the deep molecular mysteries of the cell, and which in the 1970s had become an increasingly common research tool for cell biologists. This was the electron microscope, an instrument that used electrons instead of light to illuminate tiny objects. It could magnify minuscule structures in the cell over 100,000 times. The electron microscope could even see viruses. Long strings of DNA snaking through the cell had also been photographed with electron microscopes.

As a young, aggressive graduate student on Staten Island in New York, Merz set herself the arduous task of learning how to stain cells with heavy metals and how to interpret the often bizarre images produced at such extreme magnification by the electron microscope. It took her more than two years before she began to feel comfortable using the huge machine, which in the mid-1970s usually filled a whole room. The most skillful part of the work involved the interpretation of the images. This generally took years of painstaking experience built up by trial and error. It was almost like the job of a CIA photo analyst— part art and part science. An inexperienced scientist could over-interpret some debris in a cell or else completely miss a vital clue in the electron microscope image. There was no substitute for experience. One of the first things an electron microscope practitioner learns is to always take a picture, never trust just your eyes and your memory. Always record what you see, because a photograph can be compared with another and it can also memorialize the view for publication.

When she was finally comfortable with spending eight

hours a day in a dimly lit room gazing at a green screen, Merz began obtaining scrapie samples from a colleague in England by the name of Robert Somerville. Somerville worked at the world-renowned Neuropathogenesis Unit, a division of the Agricultural and Food Research Council (AFRC) in Edinburgh, Scotland. His interest was in determining in what part of the cell the infectious scrapie particle resided.

One of the first times Merz looked at a scrapie sample on the screen she saw nothing that leaped out at her. She took a photo anyway. Later, she noticed a series of "sticks" in the sample, something she had completely missed when just looking. Slowly, over time, she saw that these sticks were only in scrapie samples and did not seem to show up in brains from healthy animals. These sticks were so tiny, she had to use magnifications of up to 70,000 to see them. It slowly dawned on Merz that the sticks, or fibrils, might be associated with scrapie. The more scrapie material she looked at, the more detailed the sticks became. She noticed they became denser in later stages of the disease, and regardless of what strains of scrapie she looked at the fibrils were always there.

Merz mused that maybe the scrapie-associated fibrils (SAF), as they were later to be called, might be proteins. The question was: Were they causing the disease, or were they being produced in response to the disease? Merz was a careful researcher and she began to teach herself the background not only of scrapie, but also of kuru and CJD. Igor Klatzo, back in the late 1950s, had observed the "plaques" in the brains from kuru patients, and had noted the similarity between the plaques found in kuru with those found in CJD. The plaques were basically bunches of protein fibrils that became knotted together. Such plaques were found in the brains of almost all Alzheimer's disease patients. The plaques in Alzheimer's disease had been analyzed and were found mostly to contain a protein called amyloid. But amyloid plaques seemed to accumulate in the tissues in response to a large number of disease states, not just Alzheimer's.

Merz wondered if there was a connection between her scrapie-associated fibrils (SAF) and the plaques that Klatzo had found in kuru brains. She began to explore the possibility that the SAF were actually the precursor to amyloid plaques, because as the disease progressed, SAFs became more and more dense in the cell. Most of the people Merz asked thought the SAFs were not amyloid plaques.

Finally, in the early 1980s, Merz and Somerville published their findings. The scrapie-associated fibrils consisted of either two or four filaments and they were morphologically dissimilar to normal brain fibrils. "However," Merz stated, "SAF do bear a resemblance to amyloid." She had insisted on putting in this sentence despite the almost unanimous disagreement of other colleagues.

It wasn't long before Merz extended her studies to humans. Cooperating with Dr. Laura Manuelidis and her husband at Yale University, Merz soon found SAF in brain tissue from CJD-infected hamsters and also in a few human brains of CJD patients. Intriguingly, Merz found SAFs in the spleens of both infected hamsters and CJD patients. Why did this finding excite Merz so much? The spleen is a roughly fist-shaped organ that lies just beneath the diaphragm. It serves as an organ that filters and monitors the blood for foreign viruses or bacteria. The spleen is packed with immune system cells. Merz's finding of SAF in the spleen was new and gave rise to the idea that maybe the infectious agent was also associated with blood, since the spleen was an organ very much associated with, and intimate with, blood and not particularly intimate with the nervous system.

The maddening and persistent question that still dominated Merz's research was: Are SAFs a cause or a consequence of disease? Inevitably, because of the high quality and originality of her work, Merz teamed up with Gibbs and Gajdusek. The three published a paper showing that SAFs were found in every species of animal inoculated with scrapie, CJD, or kuru. Thus, they were a perfect marker for this family of diseases. Gajdusek and Gibbs felt that the infectious agent were comprised of SAFs, although that had not been proven. The circumstantial evidence was becoming stronger.

The frustrating thing about this burst of research from the brilliant electron microscope work of Merz and her colleagues was that there was still an aura of mystery enshrouding the disease. Were SAFs associated with the virus? Did they contain an "unconventional slow virus" to use Gajdusek's words? What was the relationship between SAFs and disease? If they caused disease, how did they accomplish this?

Whatever the infectious agent was, Prusiner was about to give it a name. In 1982, Prusiner wrote a groundbreaking paper that sent shock waves through the science community. The shock did not originate from the scientific data so much as from his decision to actually name the novel infectious agent. Prusiner named the agent a "prion" for *pro*teinaceous *in*fectious particles.

Much later, Gajdusek would ruefully describe a conversation he had had with Prusiner in New Guinea: "I pointed out to him that I would give the disease agents a proper name when we were sure what their molecular structure was. I made this point repeatedly with him, explaining that it was premature to name them since, although we knew we had no nucleic acid, we were not sure of their biochemical nature. I had not realized that Stan would not give me the prerogative of naming them when the appropriate information was at hand. It was a clever political move on his part to jump the gun." Gajdusek was the acknowledged father of the field. According to Gajdusek, it was the unspoken rule that when the time came, Gajdusek would coin the term that would describe the infectious protein.

The reaction to Prusiner's 1982 paper was intense. "Publication of this manuscript, in which I introduced the term 'prion,'" Prusiner would recall later, "set off a firestorm. Virologists were generally incredulous and some investigators working on scrapie and CJD were irate. The term prion, derived from protein and infectious, provided a challenge to find the nucleic acid of the putative 'scrapie virus.' Should such a nucleic acid be found, then the word prion would disappear! Despite the strong convictions of many, no nucleic acid was found."

Prusiner found that the backlash to his bold move was much stronger than he had expected. Rather than attacking the data, Prusiner found many of the attacks were personal. "At times," he said, "the press became involved since the media provided the naysayers with a means to vent their frustration at not being able to find the cherished nucleic acid that they were so sure must exist. Since the press was usually unable to understand the scientific arguments and they are usually keen to write about any controversy, the personal attacks of the naysayers at times became very vicious."

Within a couple of years, Prusiner had actually chemically isolated the protein in a pure enough form to give it to master sequencer Leroy Hood. It is not an exaggeration to say that Leroy Hood has been at the forefront of practically every technical advance for the past thirty years whether in protein or gene sequencing, a process that unravels the exact order of the arrangement of the chemical-building blocks to reveal the code of DNA or provide information on protein structure. Born in Missoula, Montana, in 1938, Leroy Hood graduated from Johns Hopkins with an M.D. and from Caltech with a Ph.D. At Caltech, Hood rose quickly through the ranks, becoming a professor. In a remarkable decade and a half at Caltech, Hood brought rapid, large-scale automated technology to the sequencing of proteins and DNA.

I met Hood once at a conference on molecular biology in Keystone, Colorado, in the late 1980s. Hood's presentation was so far ahead of those of the rest of the conference presenters that it seemed to me that he had parachuted in from the future. When other researchers were talking about isolating individual genes, Hood was talking about sequencing entire organisms. There was nothing small about Leroy Hood's vision. Indeed, Hood's inventions were the key to geneticist Craig Venter's well-publicized triumph in sequencing the entire human genome a few years later, thereby leapfrogging the billion-dollar effort by the United States government. In short, when it came to the technology of sequencing either proteins or DNA, Leroy Hood was the natural person for Prusiner to approach.

With his usual skill, Hood quickly determined the sequence and structure of the unusual prion protein. In a series of brilliant papers throughout the 1980s, Prusiner, Hood, and their colleagues determined the genetic sequence of the prion protein. The well-oiled—and by now, well-funded—Prusiner lab machine was hitting its stride. At UCSF it was common knowledge that if you survived the Darwinian atmosphere in the Prusiner lab, you were probably destined for a highly productive career. It was the rule rather than the exception for all lab lights to be burning at 3 AM on a Saturday night, with teams of postdoctoral fellows working around the clock. In only a few years, Stan Prusiner's operation had come to dominate the traditionally slow-moving field of scrapie research.

Prusiner's team called the protein "prion protein" (PrP). Once the genes for the prion protein and scrapie prion protein (PrPSc) were isolated, a huge leap forward was possible. When Prusiner and company checked whether the gene for the scrapie prion protein was found in normal cells, they were dumbfounded. The gene was found in normal cells! Yet the scrapie prion protein had very different properties from the normal protein. How could a single gene encode for *both* a normal protein and a completely different infectious scrapie protein?

After much work, a crucial difference between the scrapie prion protein and the normal prion protein emerged. The normal protein was easily digested with certain enzymes, whereas the scrapie protein was resistant to this treatment. This could only mean one thing. The scrapie prion protein had a different shape. If the scrapie prion protein had a shape different from the prion protein, that explained why the same gene could encode both a normal and a diseased form of the protein. The diseased form was infectious and also caused the normal protein to change its shape into the diseased form.

This was truly revolutionary science. It brought to mind Kurt Vonnegut's famous book *Cat's Cradle*. In Vonnegut's book a single crystal of "ice nine" was capable of initiating a chain reaction that turned all water molecules into ice. When dropped into a tub of water, the single crystal of "ice nine" quickly catalyzed a chain reaction, changing molecule after molecule into a different form, eventually leading to a solid mass of ice in the bathtub. The same concept applied to the diseased scrapie prion protein when it entered a cell. The scrapie prion protein caused a domino effect inside the cell. It caused the normal-shaped prion protein to adopt the diseased shape and thus spread through the cell, eventually killing it. Needless to say, these concepts astonished scientists. They had never really seen anything like it before. The idea of an infectious protein that had no nucleic acid was simply unacceptable to many scientists, not to mention longtime scrapie researchers.

The controversy did not simmer; it boiled. Prusiner was correct that his 1982 paper provoked a backlash. Many hardworking researchers in the field felt that Prusiner had jumped the gun without sufficient data and simply guessed that the prion was a "protein only" infectious agent, a concept that was unheard of in biochemistry at that time. But by the 1990s, the term "prion" was here to stay. The media had begun using it, of course, if only to highlight the controversy between Prusiner and his colleagues.

In just over a decade of research, Stanley Prusiner had completely altered the staid, slow-moving arena of scrapie study. Correctly, he had aggressively addressed the most vexing question that haunted work in scrapie, CJD, kuru, and TME: the nature of the elusive infectious entity. It was the elusiveness and the apparent indestructibility of the agent that had made it so terrifying. In only a few decades, the mysterious agent had destroyed the fabric of Fore society in New Guinea, and had killed thousands of sheep worldwide. And finally, via Creutzfeldt-Jakob disease, the agent had killed thousands of humans. In putting together his huge research machine at UCSF, with thousands of mice and hamsters and a mere sixty-day turnaround for obtaining results, Prusiner had revolutionized this field of research. Fifteen years later, Stanley Prusiner would be awarded the Nobel Prize.

Prusiner's research was a catalyst for hundreds of scientists to leap on the prion bandwagon. The avalanche of research into the enigmatic prion protein that followed sought to discover how a protein that appeared normal in the cell could turn around and create such devastation. Beginning in 1991, it became obvious that different people carried different mutations, or changes, in their prion protein and that the combination of these genetic changes made them more or less susceptible to succumbing to CJD or to kuru. For example, careful genetic studies of Fore people who had apparently eaten kuru brains but who had *not* died of kuru showed that the survivors carried a mixture of mutations at position 129 of their prion protein (in other words, they were heterozygous) that protected them against kuru. Later, it turned out that people who carried 100 percent of a specific mutation at position 129 (in other words, they were homozygous) were more susceptible to both CJD and kuru. The lethal mutations are now known as "homozygous Met/Met 129" and "homozygous Val/Val 129." People with either of these mutations were more susceptible to dying of CJD: They had earlier onset of disease and they died earlier.

This research indicated that kuru or CJD is more likely to infect genetically susceptible people. It turns out that about 40 percent of the European and U.S. populations have the "Met/Met 129" lethal mutation and about 13 percent have the "Val/Val 129" lethal mutation and thus are more likely to succumb to and die of CJD. Of course, this is not to say that those who have a mixture of the Val or Met 129 mutations cannot contract CJD; they can. And in another twist to an already complicated story, researchers at UCSF subsequently showed that the prion protein could adopt many different shapes in the cell, not just two, and that some of these shapes were much more lethal than others.

While this worldwide burst of creative research led to a more sophisticated view on the exquisite balance between genetic susceptibility to prion disease and the complexity of the biological actions of different prion shapes within the cell, it failed to provide answers to a few basic and haunting questions. One such question was: Could the twenty-three-year-old girl named Bertha Elschker who walked unsteadily into Dr. Hans Creutzfeldt's clinic in Breslau, Poland, on June 20, 1913, have been a homozygote who had spontaneously contracted CJD—or had she eaten tainted meat? We will never know.

10

The Silencing

I t was April 1985. The owner of a mink ranch in the tiny hamlet of Stetsonville, Wisconsin, was panicked. Many of his animals had become increasingly hyper-excitable; they appeared to have difficulty chewing and swallowing, and their tails were arched over their backs like those of squirrels. Most disturbing was the fact that the mink, normally very clean animals, seemed to have lost this habit; they were leaving their droppings throughout the pen rather than in a single area. And every day dozens of these animals ended up dead.

Once again, like the latter-day Sherlock Holmes and Dr. Watson, the Wisconsin veterinarian Richard Marsh and his colleague Gaylord Hartsough were on the scene. Shortly after arriving at the ranch, Marsh and Hartsough agreed that the adult mink were showing typical signs of TME.

Throughout the 1970s, Marsh had continued working on

scrapie and TME. He found that TME was transmitted with extraordinary efficiency into hamsters with much shorter incubation periods than usual, a fact that Prusiner had been quick to pick up on. Not only did the hamsters get sick quicker, but the amount of infectious material in their brains was extremely high. And by 1980, Marsh had shown convincingly that the scrapie agent was capable of replicating in the nerve cells of the eye. The fact that infectious material was in the eye was a potentially worrying finding if the results could be replicated in humans. But even after two decades of productive research, Marsh, Hadlow, and their colleagues were still short of evidence regarding the origin of TME.

The situation took a dramatic turn during the Marsh and Hartsough investigation of the Stetsonville farm in April 1985. Over the next five months, the two researchers visited the afflicted ranch often. They necropsied dozens of animals and established that the mink had the characteristic spongiform neurons. About 4,400 mink died from TME in Stetsonville during that period. This provided the two researchers the perfect opportunity to really nail down the source of the massive outbreak.

Marsh and Hartsough spent days carefully questioning the mink rancher about what the mink had been fed. It turned out he had kept careful records. Fish supplements, ground-up "downer cows" from a fifty-mile radius of his ranch, and a small amount of horsemeat were the ingredients. The owner was conscientious. He was an old-timer who did not believe in the modern feeding practices of purchasing expensive bonemeal feeds from the rendering industry. Instead, he personally scoured the local area for downer cows and then loaded them onto his trailer and took them back to his ranch. The owner was adamant that he had fed absolutely no sheep to his mink, nor had he fed the mink any commercially rendered feed products.

When Marsh inspected the records, he was sure the mink rancher was telling the truth. Every downer cow he had ground up to feed his mink was entered carefully together with the date. The Stetsonville rancher was telling them exactly what the Canadian rancher had told them way back in 1963. No sheep were fed to the mink. The downer cows were the main ingredients of the mink's diet. The implication was chilling.

Both Marsh and Hartsough discussed the strong possibility that an undetected transmissible agent that resembled scrapie was already loose in the United States cattle population. There really was no other explanation if sheep infected with scrapie had not been near the mink. Furthermore, to date all attempts to inoculate mink with scrapie-infected sheep brain had been unsuccessful.

Alarmed by these ramifications, the normally quiet and affable Marsh attended a cattlemen's meeting in late 1985 to warn the ranchers of what he had found. "I went to the meeting of the U.S. Livestock Association," Marsh told researchers Sheldon Rampton and John Stauber,

and reported that there is strong circumstantial evidence that mink encephalopathy is caused by feeding infected dairy cows to the mink. I tried to put them on the alert to look for such a disease in dairy cows. To try to experimentally back this up, we inoculated two

Holstein steers with the mink brain in the summer of '85. Within 18 months, both of these cows came down with the brain lesions seen with spongiform encephalopathy. . . . The reason I feel so strongly is that we have a unique perspective here in this laboratory because we work on mink encephalopathy. I believe mink encephalopathy is caused by feeding downer cows to mink. I think our evidence is very sound, and I think our experimental studies on inoculation of cattle are sound.

Marsh's brief description does not do justice to the careful experiments that he had conducted. He had inoculated dairy cows with TME from minks and waited for the results. Within eighteen months, the animals began behaving strangely. Necropsies revealed the cows had spongiform neurons characteristic of scrapie, kuru, CJD, and TME. There was no mistaking the telltale signs. In another series of experiments, Marsh transferred the infectious agent back to mink by grinding up the brains from the dead cows and reinoculating mink with the material. Obligingly, the mink died of spongiform encephalopathy. Marsh had convincingly shown that the disease was transferable both ways—from mink to cows and back to mink—and that in both cases the disease was lethal.

When Marsh began to raise public warnings about his data, he was completely unprepared for the reaction he received. He had expected concern on the part of the beef and dairy industries, because if he were correct in the interpretations of his data, a time bomb was lurking beneath the facade of the images of happy, well-fed cows responsible for a large percentage of the American food supply. This message was not at all welcome and Marsh didn't have the personality needed to see it through. He preferred the quiet backwaters of the research lab that hummed with orderly scholarship. He had no inclination to seek the public stage. His decision to go public with his research was born out of concern for public safety and for the public well-being.

Marsh wasn't personally strong enough to ruthlessly protect a patch of scientific turf. The cattle industry had spent years getting to a stage where "economic" feeding practices were increasing profits spectacularly. A lot of money was at stake. Now this scientist from some university was claiming that some of their immensely profitable feeding practices were dangerous. Marsh's conclusions would be devastating to the rendering and cattle industry if they ever became topics for the mass media.

Meanwhile, the unintentional spread of prions, this time as the result of a medical intervention, had led to another disaster. Humans, rather than mink, were direct victims in this story, which began in the late 1950s during a boom time for applied biochemistry. The Medical Research Council (MRC) labs in Cambridge, England, had led the world in the purification of proteins. The first purification and characterization of insulin occurred in the 1950s, culminating in the publication of the complete structure of insulin in 1955, through the brilliant work of Fred Sanger and his team. In the late 1950s and the early 1960s, Max Perutz and his team had solved the structure of hemoglobin, the fundamental oxygen-carrying protein in the blood. Both Sanger and Perutz won Nobel prizes for their work. Insulin purified from the pancreas was used to treat diabetes. In spite of differences in the structure of insulin isolated from humans, cows, and pigs, pig and cow insulin was initially used to successfully treat human diabetics. The idea that biologically active proteins could be isolated from human tissues and then used therapeutically sparked a large wave of excitement. This was an exhilarating time for biochemistry.

Growth hormone (GH) was another protein originally isolated in the early 1950s from pituitary glands in the brain. The purified protein had some astonishing properties. Doctors found that injecting youths afflicted with pituitary dwarfism with a daily dose of growth hormone reversed their short stature and enabled them to lead normal lives. Before growth hormone was purified and characterized, children with pituitary dwarfism often lived lives as outcasts, growing to a maximum height of only four and a half feet.

The first attempts to treat children of short stature with growth hormone purified from pigs and cattle failed. Medical researchers quickly discovered that growth hormone did not have the same biological activity between species as insulin did. Only growth hormone purified from the pituitaries of primates or humans worked on humans. In 1957 Maurice Raben purified growth hormone from human pituitaries and used a variety of methods to kill bacteria and viruses during the isolation procedure. Raben then injected the material into a seventeen-year-old boy with pituitary dwarfism. After receiving injections two or three times a week, his growth increased fivefold and the boy was able to live the normal life of an adult. By 1959, once the results were published, widespread treatments in the United States and in Britain had begun.

Doctors found that GH-deficient children would grow faster, often achieving two to three times their pretreatment growth rate during the first year of therapy. This increased ("catch-up") rate waned over time, but the children sometimes continued to grow at a normal rate while receiving therapy. GH-deficient children who responded well to GH were taller as adults than they would have been if not treated.

News of the dramatic reversals of short stature spread like wildfire. Within a short period of time, youths and parents of youths from all over the world were clamoring for this wonder drug. Supplies quickly ran low. The purification of growth hormone from human cadavers began to scale up. So great was the demand that by the early 1960s thousands of pituitary glands were being pooled in a frantic attempt to keep up with the insatiable demand for human growth hormone (hGH). Thousands of children received regular growth hormone treatment and the effects were indeed miraculous. By the early 1980s, the growth hormone treatments were being touted as medical miracles and one more example of human ingenuity.

The first sign that something was wrong happened in June 1984 and was reported by Paul Brown, an acknowledged father of neuropathology diseases, in a 1988 article in the journal *Pediatrics*:

In May 1984 a young man flew from San Francisco to Atlanta en route to Maine to visit his grandparents. As he rose from his seat to change planes, he complained of dizziness. His mother, who was experienced in the diagnosis of hypoglycemia, gave him some candy and watched him closely for the rest of the trip. Nothing more happened and the incident was forgotten. Several days later in Maine, he turned down an offer to go for a spin on the lake in his grandfather's motorboat, saying "he didn't need to go for a spin because he was already dizzy." On his return from Maine, the patient went back to school but his dizziness persisted and now his speech seemed slightly changed. When we went to his physician they knew him because he had received regular injections of hGH since he had been two years old. As the condition worsened, and obvious neuropathology began to manifest, some of the attending physicians considered a diagnosis of CJD, but quickly rejected it because of the young age of the patient. The young age of patients coming down with CJD symptoms was to become epidemic in the 1980s. After the first patient's death, the diagnosis of CJD was confirmed because his brain had the characteristic spongy holes throughout his neurons.

By late 1985, three youths had come down with CJD. The cases received a lot of attention because the patients were so young. Then it was realized that each of the victims had received growth hormone injections several years previously. Dozens of deaths occurred worldwide in the following years. Suddenly, the chances of an hGH recipient becoming a victim of CJD was reduced from one in a million to one in a hundred. Pandemonium broke out.

Between 1963 and 1985, about 7,700 people received

human growth hormone made by the National Hormone and Pituitary Program (NHPP). Twenty-six of them got CJD.

People treated with hGH in other countries also got CJD. In France, of the 1,700 people treated with hGH, 89 people got CJD. In England, of 1,848 people treated, 38 people got CJD. In New Zealand, five people got CJD. Two people got CJD in Holland. One person got CJD in Brazil, and one in Australia. The New Zealand patients and the patient from Brazil received hormone made in the United States, but it was not identical to the hormone distributed by the NHPP. France, Britain, Holland, and Australia produced their own hormone.

The growth hormone tragedy was eerily similar to the widespread inoculation with scrapie during the attempt to create a vaccine for louping ill at Compton. The spread of CJD was an excellent example of a successful transmissibility experiment. The tragic episode demonstrated that CJD could easily be transmitted to humans simply by giving them injections of small quantities of CJD via the pituitary glands from cadavers. The growth hormone fiasco showed that even if only one in a million people had CJD, you could dramatically increase the odds of spreading the disease by pooling millions of pituitary glands from cadavers. Luckily, by the 1980s, when the problem was seen in all its ramifications, the genetic engineering revolution was in full stride. Instead of using the laborious process of pooling millions of pituitary glands from cadavers, it was necessary simply to isolate the gene sequence for hGH and insert it into bacteria and grow huge quantities of the protein.

This unfortunate episode not only highlighted the infectiousness of prions, but also added to the number of fatal mistakes that had been made over the years involving this spreading epidemic. This human-to-human prion spread is another example of a continuum of disasters that began with the massive inoculation experiments at Patuxent. The other end of the continuum is nowhere in sight.

The next horror was stirring in the United Kingdom. But nobody knew about it yet.

11

Mad Cows

Tom Forsyth, the manager, and his head dairy stockman, John Green, were the first to note the unusual symptoms in their dairy herd at Plurenden Manor farm near High Halden, Kent. Located in southeastern England, Kent is full of the rolling hills and lush pastures that Holsteins and other breeds have turned into milk for hundreds of years.

In April 1985 a cow named Jonquil had started behaving oddly. "From being a nice quiet cow," John Green told *The Times* of London, Jonquil "had turned into a nuisance in the milk parlour, acting aggressively towards the other cows. She seemed to hallucinate." Green was in charge of the herd of 300 Holstein Friesians at the farm; Forsyth's herd had an excellent reputation.

Reflecting over the incident a decade later, Forsyth told *The Times*: "Looking back over the years since then, horror is the only word to describe my feelings; horror that we had got something that seemed to be out of control. We did not know where it was coming from and we did not know how to put it right. Even now the origin of the disease is still not known for certain."

Forsyth and Green first thought that the cow might be suffering from "grass staggers." This common condition can affect cattle after they are let loose into the pasture in the spring. Caused by a shortage of magnesium in the blood, it is characterized by shivering and staggering. Green called in the local veterinarian, Colin Whitaker, who found that Jonquil had cystic ovaries, a common cause of aggression in these cows. "I treated the ovaries, which got better," noted Whitaker, "but the cow did not. She got worse and was very unsteady on her feet. I thought she might have a brain tumour or abscess." After treating her ovaries, Whitaker also began giving the animal magnesium supplements in case her condition was a particularly virulent form of the grass staggers. The supplements had no effect on Jonquil's condition.

As April turned into May, Jonquil's behavior became increasingly bizarre and her balance deteriorated rapidly. She would sometimes react violently to some unseen stimulus. Sudden sounds set off avalanches of shivering and trembling. It was obvious she had no idea of her surroundings most of the time. She twitched continuously. When she could no longer stand, they made the humane decision to put her out of her misery. Jonquil was hauled off to the local rendering plant, where her carcass was taken apart and made into feed for other animals.

Forsyth quickly forgot about Jonquil, thinking she had

died of some mysterious and unexplained ailment that seemed to crop up in his herd every few years. As long as it was only one animal, he was not going to worry. Because Holsteins need to be milked twice a day, it is routine for dairy farmers in England to know each animal intimately and to learn to recognize their individual "personalities." Any deviation from expected patterns is noted quickly.

Then, at the beginning of 1986, Green and Forsyth noticed several other animals in their herd were not cooperating and were becoming more aggressive. They seemed nervous, bumping into other cows and becoming increasingly difficult to handle. The animals' behavior seemed eerily reminiscent of Jonquil's deterioration. Forsyth realized he was dealing with something novel and frightening. "With our vet," he would later tell the reporter from *The Times*, "we considered a whole range of possible causes, from lead poisoning to rabies, but nothing made sense."

Still other cases began to appear, always with the same symptoms. Their behavior would become aggressive. Then the cows would begin to stagger and a few weeks or months later they were dead. After the Plurenden Manor farm incidents, subsequent cases appeared in Cornwall, Devon, and Somerset, three other counties in southeastern England. It was then that the telephones began to ring and the farmers began to chatter in the dairy farming community. Something was wrong, and because it appeared in three different counties simultaneously, it did not seem to conform to an infectious disease.

After all the creativity of the local veterinarians had been exhausted, the Central Veterinary Laboratory (CVL) of England's Ministry of Agriculture, Fisheries and Food (MAFF), now the Department for Environment, Food and Rural Affairs, became involved. In 1986, the ministry launched a low-key investigation. Because of the number of new cases coming in, they already knew that this was not a business-as-usual toxicity investigation. It was a measure of how unprepared the British ministry and the veterinarian community were that it was months before an adequately competent necropsy procedure, complete with properly fixed brain, was completed. In order to stop the rapid deterioration that occurs in the brain beginning immediately after death, it is necessary to immerse the brain, or slices of the brain, in formaldehyde solution. If the practitioner waits too long after the necropsy before getting the tissue into formaldehyde, deterioration is rapid. Alternatively, too little formaldehyde solution will not completely fix the brain.

A veterinary pathologist by the name of Gerald Wells, working at the CVL, began to perform pathology examinations on brain tissue from some of the sick animals. There were multiple false starts and he complained of looking at badly fixed brains or tissue obtained from incompetent necropsies. Finally, in November 1986, as Wells looked under his microscope, he found his field of view filled with neurons that were riddled with holes. Wells immediately thought "scrapie." The cow's brain had the now familiar Swiss cheese–like brain pathology pattern of sheep with scrapie. But Wells knew scrapie had never before been described in cows. This was something new.

Wells quickly penned the first scientific article describing this new threat to British agriculture. He called it "bovine spongiform encephalopathy," or BSE for short. But in the slow, sedate world of veterinary research, it did not appear for almost a year after he submitted it. "Previously healthy cattle, in good bodily condition," he wrote in the *Veterinary Record*, "became apprehensive, hyperaesthetic and developed mild incoordination of gait. Their mental status was progressively altered and normal handling procedures evoked kicking. Fear and aggressive behavior were recorded and auditory stimuli produced exaggerated responses, even falling. . . . Eventually frenzied behavior and unpredictability in handling, or recumbency, necessitated slaughter."

Wells's scientific article recorded the telltale spongy appearance of neurons and the parallels with scrapie. The article did not make huge waves in the media and the parallels with scrapie served merely to highlight a curiosity of veterinary science rather than an accelerating, catastrophic epidemic. Thus, the initial response to the Wells article was muted. That certainly was not the case behind closed doors at the ministry.

Immediately following Wells's discovery at the end of 1986, there was alarm at the ministry. "I'll remember it till my dying day," recalled Keith Meldrum, the chief vet at the ministry, for Peter Martin of the *Mail on Sunday Magazine* a decade later. "I was just down the corridor when the guys from the Central Veterinary Laboratory came in. Quite a hubbub . . . they were talking about scrapie. I understood scrapie. But they were also talking about things I'd never heard of—Creutzfeldt-Jakob disease and some thing called 'kuru,' a rare form of CJD once common in New Guinea among the Fore tribe of cannibals."

Meldrum and most of the vets at the ministry had never heard of Carleton Gajdusek, Joe Smadel, or any of the pioneers who had brought the human dimensions of this disease to the public. But they were soon to redress this shortcoming. In 1987 the number of cases began accelerating and they were coming in from all over England and Wales, sometimes at the rate of more than a dozen a month. By the end of 1987, there had been about 420 confirmed cases.

In June 1987 the British government quietly established a commission led by Dr. John Wilesmith, a veterinarian who worked with the ministry, to investigate the cases. The year before, Wilesmith, who was well acquainted with scrapie and how it spread, had been appointed head of the epidemiology department of the CVL, which reported to the ministry. The director of the CVL asked Wilesmith to investigate the epidemiology of BSE and was told to discuss it with his colleague, Gerald Wells of the Pathology Department. Wells, of course, was the pathologist who had seen the original chilling microscopic vista of spongiform lesions in the cow's brain and had given the disease its now well-known initials, BSE.

Wilesmith's expertise was quickly put into action. On June 9, he personally visited several affected farms in Somerset. Throughout the summer months, he visited dozens of stricken dairy farms and obtained reams of data on the genetics, feeding practices, and other details.

Wilesmith quickly discovered that the widespread use of organophosphate (OP) insecticides was common to each of the farms that he visited, but he equally quickly ruled out these chemicals because the lesions of the diseased cows differed markedly from anything that had been observed as a result of OP poisoning. By late August 1987 Wilesmith's team had acquired intensive data on fifteen affected herds and had effectively ruled out insecticides and other pharmaceutical treatment of animals. Feed, however, required more work. But by December 1987, the evidence had suggested to Wilesmith that BSE was associated with a feedborne source, which was most likely to be the mix of ground-up meat and bonemeal that was fed to the animals.

With remarkable efficiency Wilesmith's epidemiologic team had quickly eliminated the possibility that causes other than feeding the cattle ground-up material from animals were responsible. After reviewing Wilesmith's data and methodology, another consultant who worked with the ministry remarked: "It was a beautiful, I would say, brilliant piece of classic epidemiology. It established that BSE is associated with the feeding of meat and bone meal. . . . That is one of the best documented evidence we have. I really cannot emphasize it too strongly."

Wilesmith's study had also noted the preponderance of BSE in dairy cows over beef cows. British dairy Holsteins produce on average about forty pounds of milk every day and milk is high in protein. In order to achieve this huge milk production, farmers had been supplementing their animals' diets with protein for decades. Although soybean was a popular source of extra protein, as farming methods incrementally became more mechanized and focused increasingly on costs, the expense of importing soybeans to Britain was deemed unacceptable. An alternative had to be found.

The British used the term "knacker's yard" to describe their rendering plants. Beginning during World War II, and accelerating through the 1960s, British farmers increasingly used the meat and bonemeal that the knacker's yard, produced as a cheap protein supplement for their dairy animals. They fed these meat and bonemeal supplements to the dairy cows twice a day. Sometimes, in order to "beef up" their beef, they used meat and bonemeal in the latter stages of the feeding process before the beef went to market. But mostly, the concentration of meat and bonemeal was in the feed of dairy cows. Wilesmith's brilliant epidemiology team had found the answer to why the disease was overwhelmingly ravaging dairy and not beef cows.

A question remained, however. It was well known that meat and bonemeal had been fed to dairy cows since the late 1940s, so why was there this very sudden eruption of BSE in multiple separate counties in the country? Wilesmith and his epidemiologists set to work to answer that question. When had the epidemic actually begun? Was there an event in the feeding practice that could be traced to the sudden emergence over a few months of this deadly new disease?

Gathering all of the data together and estimating an incubation of approximately four years for BSE, they projected that BSE had entered the dairy cow system around the winter of 1981–1982. Three of Wilesmith's team had good knowledge of the British system of knacker's yards. They set about an exhaustive survey of the rendering process, focusing on whether anything had changed in the early 1980s. In Britain in the mid-1980s there were some forty-five knacker's yards in production.

The public generally avoided knacker's yards because of the smell. Within several blocks of the knacker's yard a foul mixture of burnt flesh and rotting meat mixed in with a definite fecal smell assails your nostrils. The overpowering stench quickly discourages most people from getting any closer to these establishments. Inside the knacker's yard, body parts from a bewildering variety of sources are cooked and chemically processed, and fats are removed. Meat and bonemeal is the end result of a complex process involving the use of chemical solvents geared to lowering the fat content of the protein supplement.

Wilesmith's veterinarians quickly discovered that an apparently unrelated event had been an important catalyst in changing the way the rendering industry operated. This change had gone unnoticed. On Saturday, June 1, 1974, Britain experienced its worst-ever industrial disaster at the Nypro works in Lincolnshire. Fortunately, because it was the weekend, very few of the 550-strong workforce were there at the time of the explosion; otherwise, the loss of life would have been immense. Twenty-nine people died and more than 100 were injured, some seriously. Almost 2,000 homes and businesses were damaged by the blast. Only eight of the bodies were ever recovered. The remainder of the workers had been carbonized by the force of the blast and by the horrendous fire that followed. People living thirty miles away believed they were under nuclear attack as a mushroom-shaped cloud rose into the sky. The Nypro plant manufactured caprolactam, an ingredient used in nylon manufacture, and had other industrial highly flammable solvents such as benzene and cyclohexane in huge quantities on site. In the wake of this huge tragedy, the British government introduced a series of extensive safety procedures in the handling of inflammable solvents.

The sudden introduction of new safety standards for handling and storing solvents forced the rendering industry to make a choice: Either spend a lot of money upgrading their facilities to accommodate the new safety regulations or simply abandon the use of solvents altogether. It does not take a genius to guess which option the industry took. Beginning in the late 1970s and accelerating into the 1980s, rendering plants simply stopped using solvent extractions in processing the wastes that ended up as meat and bonemeal. This sudden and dramatic change in the rendering process meant that many ingredients that previously had been extracted and removed by the solvent treatment now went straight into the meat and bonemeal product.

Secondly, also for economic reasons, around the late 1970s, the majority of rendering plants had begun to adopt a new continuous processing methodology from the United States that superseded their traditional batch processing. This new system increased the tonnage of animal parts flowing through the system and so increased profits.

Wilesmith's detectives had now outlined two very substantial changes to the production of meat and bonemeal that had been simultaneously adopted at exactly the correct time period and both of which contributed to the transfer and survival of the deadly infectious agent in the meat and bonemeal product.

Wells's paper describing BSE had been published in October 1987, and two months later, Wilesmith's data were pretty much in hand at the ministry. Their internal response had been lightning fast. The stage appeared to be set for the British government to now react rapidly to this news by informing the public and to quickly institute widespread changes in the practice of feeding animals the ground-up remains of other dead animals.

Wrong.

The public, including the media, remained blissfully unaware of what was happening in the idyllic British countryside. The ministry, it turned out, had no intention of informing the public. In late December 1987, only a single article had been published in the British media about the disease, and the tone of the article described a new curiosity in cattle that bore some resemblance to scrapie in sheep and to an unknown disease called kuru. *The Times* article also quoted a British veterinary official, Dr. Tony Andrews, "who did not yet see BSE as a serious threat to cattle health." By the time *The Times* article was published, over 400 cases had been reported and the number was climbing every month. The ministry continued the policy of secrecy even as the number of new cases mounted.

The minutes of an April 14, 1988, government ministers' meeting held to assess the urgency of the BSE crisis warned that "this issue [the BSE crisis] could assume a higher profile in the immediate future." The meeting concluded: "After some discussion it was agreed that this could be handled in a low profile way . . . a written PQ [Parliamentary Question] was not appropriate, but a paragraph should be included in *The Veterinary Record.*" Reading between the lines of the dry bureaucratic language, we can easily ascertain that the ministry had no intention of spilling the beans to the public that hundreds of cows were dying of disease. They vetoed any suggestion that questions could be raised in Parliament concerning BSE and instead suggested putting a paragraph into an obscure veterinary journal that nobody read. And the language in the government article in the *Veterinary Record* stated: "BSE must be seen in perspective. The number of confirmed cases (455) is very small compared with the total cattle population of 13 million. The number of cases is expected to increase, but if, as anticipated, it behaves like similar diseases in other species, only small numbers of incidents relative to the total number of cattle incidents are likely to occur."

As if to back up a general ministry mandate to minimize the threat, the minutes of the April 14 ministers' meeting continued: "It was agreed that our public line should be that we were taking the issue seriously, that a number of steps were already under way: but that the question must be kept in perspective."

The secrecy continued until a one-two punch from the *Sunday Telegraph* and a front-page story in *Farming News* on April 22, 1988, broke the whole story wide open. "A mystery disease which riddles cows brains with holes and drives them mad is to be investigated by the government," stated the *Telegraph*. "There are fears that people might catch the disease for which there is no known cure. It has been identified only in Britain where there have been 421 confirmed cases and sick animals which do not die are destroyed." Finally, the disease was out in the open.

The *Sunday Telegraph* and a couple of low-key government statements about BSE created a stirring of unease throughout Britain. The *Telegraph* article went further by criticizing the government's lack of a sense of urgency in dealing with this new threat: "The Ministry's low key approach has been strongly criticized by some vets. Mr. Tony Andrews, a senior lecturer at the Royal Veterinary College, said yesterday, 'The Ministry has to come clean about this disease. We simply don't know if it is a danger to humans.'"

This was the same Tony Andrews who, only a few months previously, had issued the bland statement in *The Times* article, stating that BSE did not yet appear to pose any danger to cattle health. What had changed in the intervening months to prompt Dr. Andrews to be so harshly critical of the government's foot-dragging? The answer, presumably: Dr. Andrews had seen the data.

In a far more aggressive article in *Farming News*, Laurena Cahill wrote: "The Ministry of Agriculture has been accused of seriously underestimating the extent of bovine spongiform. The charge by vets up and down the country is made more acute by fears that the brain disease could be transmitted to humans." The newly converted Tony Andrews again weighed in: "It's time the ministry came clean about the disease. There is a distinct feeling that the investigation center at Weymouth is over-secretive about its findings."

The ministry was outraged. In a remarkable secret memo that begins "This is just the article we do not want," the memo goes on to speculate about a "mole" and then deals with Dr. Tony Andrews: "Tony Andrews if properly quoted is becoming even more of a menace . . ."

Throughout the first stages of the BSE crisis it appears from the memoranda that flowed between individuals and departments at the agriculture ministry, that the ministry's primary focus was not public safety. Rather, it was to keep the whole thing secret. In spite of a superbly executed internal response orchestrated by Dr. Wilesmith's team in rapidly homing in on the meat and bonemeal product as the culprit, the ministry appeared to degenerate into an alarmed paralysis. Through a combination of bureaucratic hubris and reluctance to rock the boat, the ministry simply did not act on the superb science that Wilesmith's team had produced.

In another classic response to a crisis, the government reacted as any government would: it created a "blue ribbon" commission to examine the problem. The group was to be chaired by Sir Richard Southwood, a famous professor of zoology at Oxford University. Sir Richard himself, although an eminent scientist, was an entomologist and naturalist, with little or no experience in neurology or pathology. The other three members of the committee were well-respected scientists, but all were retired from active practice and none was an expert in transmissible spongiform encephalopathy. In announcing the existence of the committee, the ministry sent a series of talking points with the press release so that if any awkward questions were asked by the press, bland, reassuring answers could be supplied. The talking points emphasized giving the minimum of information possible. One example:

"Q: Where has the disorder been identified and how many cases confirmed?"

"A: In most parts of the country but with a preponderance in the south. (If pressed about the number of cases, say that 421 cases confirmed on 352 farms.)"

Even Sir Richard Southwood himself was concerned about the casual and informal terms of reference of the group that the ministry wanted to create. The ministry appeared to want a group behind the scenes that they could trot out to the press if the pressure became too great, but that otherwise would remain a low-key, informal, and probably nonproductive committee. Southwood went along with the ministry's insistence on keeping all the findings of his committee secret, but he insisted that the group at least adopt the formal name of "working group" or "committee." Initially, the ministry was reluctant to even formalize the Southwood group as a committee, because they feared it would give the impression that the ministry was worried about the disease.

On May 3, 1988, Dr. Wilesmith presented a summary of his findings to date in an internal memo. In it, he homed in on sheep and scrapie. "The sheep population [in Britain] has increased markedly since 1982," stated the memo, and "the practice of skinning sheep's heads and harvesting of sheep brains has almost ceased resulting in more unsplit sheep's heads going for rendering." Finally the memo stated that "there appears to have been an increase in the incidence of scrapie." Wilesmith also noted that "low temperature renderers have commenced operations."

This internal memo, which was marked "in confidence," is crucial to understanding where the ministry was heading in their interpretation of the origin of BSE. They were near certain in their conclusion that BSE had originated from the rendering of scrapie-infected sheep. Wilesmith may have been unaware of Richard Marsh's groundbreaking work on the Stetsonville, Wisconsin, outbreak in mink in 1985. Marsh's work strongly implied that the infectious agent was in cattle. There were two possibile origins for the BSE epidemic in Britain that remains, even in 2004, a puzzle. Did BSE originate in sheep or in cows? Wilesmith, who was the driving force in the ministry's behind-the-scenes rush to obtain data on BSE, seems to have rejected the idea that BSE might have originated from cows. Therefore, Wilesmith's internal memo marked the creation of policies that only considered a single origin of BSE: sheep. And Wilesmith's memo was written fully three years after Jonquil's first symptoms appeared.

By the time the Wilesmith memo had circulated throughout the bureaucratic levels at the ministry, the British public remained blissfully unaware of the huge, terrifying disaster that was bearing down on them like a freight train.

12

Cover-up

Everybody knew that scrapie had been around for a couple of hundred years and that as far as anybody knew, no humans had ever died from eating scrapie-infected meat. The British Ministry of Agriculture's decision to treat BSE like scrapie meant protecting the animals from the disease. The ministry simply ignored the possibility that this disease could ever be transmitted to humans.

On June 2, 1988, ministry officials met with representatives of the United Kingdom Renderers' Association (UKRA) to discuss the next steps. A private and confidential report on the meeting states that the ministry had decided to make BSE "a notifiable disease" and suspended the sale and use of animal proteins in feeds for cows and sheep. The renderers were not pleased. They felt that this drastic action "was a very serious blow to the industry," and that the small operators would find it hardest to meet with "the likely

requirements." The UKRA representatives viewed the measures as draconian and stated that they would have difficulty
complying with them.

Nothing was decided about what would happen to the
thousands of tons of meat and bonemeal product that had
already been manufactured, although the ministry order gave
the rendering industry twenty-one days to stop the flow of
meat and bonemeal through the system. Many dairy farmers
had already spent hundreds or even thousands of pounds to
purchase the meat and bonemeal in bulk for their animals.
Were they simply to burn their investment? And if they
were, who was going to enforce this destruction? Nothing
was mentioned of the fifty-eight renderer plants around the
country that also had huge stockpiles of meat and bonemeal.
The renderers proposed that they voluntarily police themselves, but the ministry overruled that option. The renderers
were concerned that if another cause for BSE were found, the
draconian measures would have needlessly hurt the industry.
The steps were already in place for an exercise in foot-
dragging that would delay the implementation of the ministry's ruling.

On the same day that the ministry met in private with the
rendering industry, an innocuous article appeared in *The Independent* newspaper in Britain. "Legislation is being rushed
through by the government in an attempt to combat a new
disease which has led to the death of more than 500 cattle,"
the article announced. The "new disease" was over three years
old, but due to the secrecy at the ministry, neither the media
nor the public knew much about it.

The next day the renderers' association and related feed

organizations circulated a memo to their members warning that big changes were coming down the pike, and that there was really nothing that could be done. They stated that the ministry would hopefully lift the ban on animal protein feeds by the end of year. This statement of hope seems a trifle quaint given that most of the public had little idea of what was truly going on. It probably reflected the ministry's success in downplaying the crisis. Aware of the muted public reaction and the absence of press reports on the issue, the renderers were understandably puzzled about what they considered an overreaction on the ministry's part. After all, so what if another scrapie outbreak was on their doorstep?

But that day, June 3, a hard-hitting article by Dr. Tim Holt from St. James Hospital, London, appeared in the *British Medical Journal* (*BMJ*). It began by noting how the tiny number of press announcements the previous year of an outbreak of brain disease in the cattle of southwest Britain "were received with alarming indifference by the medical profession as well as by the general public." Holt warned that BSE was akin to kuru, scrapie, and to CJD, and that the spongiform diseases were always fatal. He also warned about loopholes in the law and that brains from these infected cattle were being sold routinely over-the-counter in butcher shops, and more chillingly, that it was not illegal for brain from infected cattle to end up in meat pies or canteen food. Unlike in the United States, meat pies are very popular in Britain and are avidly consumed, especially by the less wealthy segments of society.

Much to the chagrin of the ministry, Britain's *Today* newspaper picked up on Holt's hard-hitting warning and gave it an appropriately sensational headline: "Danger of Killer Meat in Beef Pies." The newspaper article stated: "A mystery bug which has killed hundreds of cattle could be passed onto humans . . . it was thought that the animals' slaughter removed any risks to humans but Dr. Holt says this view is 'naive, uninformed and disastrous.'" The *Today* article also carried the ministry's standard bland rebuttal: "[the] spokesman said some affected animals might have entered the food chain 'but there is no evidence of any risk to humans.'" The ministry's dismissal ignored Holt's contention that the present regulations were so full of holes you could drive a meat truck through them.

The ministry also ignored the insistence by Holt that BSE was essentially undetectable in the brain of a cow until the animal reached the terminal stages of disease and began to stagger. At that stage, the cow was approaching death. It was also obvious from Holt's article that if cattle reached the slaughterhouse before showing any of these symptoms, they would be slaughtered as usual, infected meat would enter the food chain, and nobody would be any the wiser. Holt's warning fell on deaf ears at the ministry in their public pronouncements.

But behind the scenes, ministry officials were alarmed. In a confidential letter to Sir Richard Southwood immediately following the publication of Holt's article, an official summarized the feed ban that was going into effect and succinctly expressed the ministry's position on BSE: "Although we are taking these steps for animal health reasons, it may help to allay some of the public's fears expressed (notably in a recent *BMJ* article which I enclose)." In other words, the public fears were simply hysterical and the ministry should just focus on protecting the animals.

While maintaining a posture of outward calm and down-playing the dangers posed by BSE, the ministry's language was sounding increasingly urgent. At a confidential June 13 meeting attended by members of the rendering industry, ministry officials described the situation using the eight letter "e" word: "The Ministry concluded they were facing an extended source epidemic. The outbreaks were occurring at a steadily increasing rate. The hypothesis was that there had been a species jump from sheep to cattle." This was the first time the word *epidemic* had appeared and it was plain that the situation, according to the ministry's internal documents, was beginning to spin out of control.

Even though the public heard much about the Southwood "working group," it seemed to wield little power. Its members performed their investigative duties admirably, but they had no power to compel the government to act. The South-wood committee met for the first time on June 20, 1988. Their first recommendation was that infected animals be slaughtered and their carcasses destroyed.

At the time of the announcement, more than 600 cows had already died from BSE on more than 400 farms scattered throughout England and Wales. Two weeks went by before the government took action on the Southwood recommendation and agriculture minister John McGregor ordered the compulsory slaughter of infected animals and their carcasses destroyed. The measure was too little too late. Three years and three months had elapsed between Jonquil's unsteadiness and the first government response that could be called decisive.

For cattle farmers, the minister's announcement bore a nasty coda. "Compensation" for the infected animals that had to be destroyed, said McGregor, "will be payable at 50 per cent of market value subject to a ceiling." This miserly level of compensation drew an immediate uproar from farmers. If the ministry paid at 75 percent of market value for cows infected with tuberculosis, why was the BSE compensation so low? Some later speculated that the numbers would rise to a huge level and the British Treasury knew ahead of time to limit its losses.

Meanwhile, behind the scenes, another remarkable charade was playing itself out. In June 1988, Dr. Richard Marsh had contacted the Ministry of Agriculture offering to help in any way he could. He offered to conduct transmissibility experiments on the British BSE material in his labs in Wisconsin. As one of the world's experts on scrapie and TME, Marsh was in a position to contribute his decades of research experience to the British crisis. But instructions on how the ministry should reply to Marsh's offer were curt and to the point: "We appreciate Professor Marsh's interest and offer of assistance and he should of course be thanked for this. For the moment however we should not offer material to him and indicate we are in the process of initiating an experiment here."

In spite of floundering for over three years, the ministry was not about to allow an expert into the equation. Perhaps one reason for this blunt refusal to allow Marsh access was that the ministry might not have enjoyed a reexamination of the basic premise behind their routine pronouncements that BSE had jumped species from sheep to cows. Perhaps Marsh would find a different interpretation. This in turn might lead to questioning about the ministry's policy on feed. But in the wake of the ministry's refusal to countenance help from Dr. Marsh on the transmissibility experiments, a flurry of memos began circulating around the ministry looking for the appropriate facilities to house mink for their own experiments. The memos also searched for the appropriate expertise in inoculating mink with BSE. If Dr. Marsh had been allowed to spend two weeks at the ministry labs, all of these problems would have been solved. Instead, the ministry created a situation where all of the pitfalls surrounding successful inoculations that had been garnered from Dr. Marsh's work were simply brushed aside. The wheel was about to be reinvented.

With another offer of help, Dr. Tim Holt, the author of the hard-hitting *BMJ* article, visited the ministry to see the facilities and to remonstrate about why they were not taking this disease more seriously. A July 27 memo, which summarized his visit, described Dr. Holt as "an earnest young man with visions of hospitals overflowing with CJD patients." The snide tone of the description was especially ironic given what lurked just beneath the horizon in Britain.

The ministry's order to destroy all infected carcasses begged the question of how exactly to dispose of the carcasses. Should they be burned? Their confusion over the issue is all too apparent in an August 9, 1988, memo from the Institute of Animal Health to Dr. Wilesmith, the epidemiologist at the ministry who had successfully "nailed" the meat and bonemeal feed connection: ". . . regarding lime pit disposal of BSE carcasses . . . we know of no data of the effect of calcium oxide [lime] on scrapie. . . . Because it would be completely unrealistic to draw conclusions for the whole car-case [*sic*] situation from the in vitro work, we are considering looking at a simulated lime-pit situation using scrapie mouse carcases [*sic*]." The idea of creating a simulated lime pit in the lab and monitoring a couple of scrapie-infected mice carcasses, with the hope of rapidly formulating a nationwide policy to dispose of thousands of BSE-infected carcasses, seems almost comical.

The "lime-pit" and "mink" internal memoranda gave the strong impression of an agriculture department under siege. In the face of a mounting crisis, the ministry seemed more obsessed with stopping internal leaks than with directly confronting the problem of a rapidly escalating epidemic. On top of this, the ministry continued to stubbornly refuse to reach out to experts like Dr. Marsh for help. It's certainly not the case that the ministry was expert on the disease. Sir Richard Southwood's letter dated September 8, 1988, reaching out to brain-imaging specialists at Oxford, reflects just how little they knew: "As you may have seen from the [news]paper, my colleagues and I have made various recommendations based, I have to admit, largely on guesswork and drawing parallels from existing knowledge of scrapie and CJ disease."

That Southwood would admit that his committee's recommendations were based largely on guesswork was remarkable enough, but it also raised the question of the reliability of the committee's subsequent recommendations. This was the same "working party" that the British people, becoming increasingly anxious as the media slowly began to wake up in the latter months of 1988, were relying on to provide the scientific backing for the policies that were emanating from the

Ministry of Agriculture. Of course, Sir Richard was quite correct in admitting that his recommendations were based on guesswork.

As 1988 wore on, the ministry conducted a secret study on the technology that the rendering industry was using to "inactivate" the scrapie or BSE agent. The study included a ministry official visiting multiple rendering plants in different locations during the months of August and September 1988. The aim was to verify how close the rendering plants were to using an autoclave in their operations. An autoclave is an instrument used in dentists' offices and in all laboratories to sterilize surgical instruments and lab equipment using prolonged exposure to boiling steam and high heat. This treatment has been used for decades to destroy all known viruses and bacteria. The ministry's hope was that, if rendering plants exceeded the performance of an autoclave, then perhaps they were killing at least a part of the scrapie/BSE agent.

In late September 1988, a ministry official internally released the disturbing news about the rendering plants operation. "I conclude," he wrote, "that at no stretch of the imagination are any of the plants I have seen analogous to the conditions pertaining to an autoclave. They operate either with moist heat below 100°C or with dry heat above it or a combination of both. My conclusion is that there is no reason to believe that they will inactivate the scrapie agent as they are and that nothing can be done to them to achieve conditions analogous to an autoclave short of replacing the processes by pressure cooking systems which alone will be analogous to an autoclave." The dry bureaucratic language hid the damning facts. The rendering plants were not operating at anything like the capacity needed to destroy the scrapie or BSE agent. It was even doubtful that they were operating at the capacity to kill bacteria or viruses. In time-honored tradition, the result of this ministry study was not released to the public.

As October wore on, seventy new cases of BSE were being reported every month and the total number of cases had climbed over the psychologically important 1,000 mark. There could be no doubt in the minds of the British that they were dealing with a fast-moving epidemic that was escalating from week to week. The basic facts had been known for well over a year due to the rapidly completed epidemiological study managed by Dr. Wilesmith back in 1987. Over twelve months had gone by, and instead of acting swiftly, the ministry and the British government had stood by and watched all of Dr. Wilesmith's epidemiological data confirm themselves over and over, month after month. The phrase "too little too late" had been used in early 1988 to describe the ministry's ponderous reaction to the unfolding epidemic. By October, "too little too late" was beginning to seem like a gross understatement of reality.

A breakthrough occurred that month when the *New Scientist* reported that BSE had been successfully transmitted to mice. BSE joined the ranks of kuru, CJD, scrapie, and TME in transmissibility. While of scientific interest, the finding had no effect on the unfolding epidemic. Historically, the *New Scientist* piece, entitled "Mice Catch Cow Madness," was perhaps the first occasion that the words *cow* and *mad* were juxtaposed to describe BSE. It was a harbinger of the popular

term "mad cow" that would deluge all later reporting on the
disease over the next couple of decades.

The safety concerns began to escalate in the fall of 1988
when Australia and Israel banned the importation of British
beef. People at the ministry began to realize that they were
fighting a losing battle. The epidemic showed no signs of
abating and other countries were becoming more alarmed at
the lack of progress against this potentially deadly threat.

As the epidemic continued, the pharmaceutical industry
also became alarmed. Bovine insulin and many other blood
products, including bovine serum albumin, were a large
profit source, and all indications were that the ministry
might move to restrict their preparation, or at least mandate
a different way of preparing insulin, albumin, and other
blood products that originated from cows. Any change in
standard procedures would automatically mean a reduction
in pharmaceutical profits, so the ministry held off on man-
dating any changes until more was known. Again, public
safety receded as the primary concern.

As word of the transmissibility of this new disease spread
through the scientific community, the ministry was besieged
with a large number of requests for samples of BSE-infected
brains. Many researchers in the United States particularly
were beginning studies on this mysterious family of diseases
and this trend had accelerated considerably since Carleton
Gajdusek's high-profile acceptance of the Nobel Prize in
1976 for his discovery of kuru. Gerald Wells, the veterinary
pathologist who had originally discovered the BSE-scrapie
similarity, was at the receiving end of a torrent of requests for
BSE samples. One of those requests came from the ambitious

biochemist Stanley Prusiner at the University of California–San Francisco.

At the time, Prusiner was becoming well known for his research on the biochemistry of the infectious agents in scrapie and CJD. In September 1988, Prusiner wrote to Wells to see if he could obtain some BSE samples. Midway through his letter requesting brain samples, Prusiner asserted: "I cannot understand why this would be a problem since more than 1,000 cows have been diagnosed with this disease and Jim Hope tells me that more than 70 new cases per week are being reported."

Prusiner was not going to accept any of the well-known British hesitancy, so he was being blunt. "If this is a problem at the administrative level," Prusiner continued, "please let me know because I have communicated in the past with a number of high government officials in British science. I would be glad to discuss this with them to facilitate the transfer of this tissue to us so that we might carry out studies that in the end would be useful to you and the British people." Prusiner's letter was not designed to ingratiate himself with Wells, but merely to achieve results.

Wells dutifully passed the letter up the chain of command. It was not long before the ministry's reply came back to Wells: "I suggest that our joint collaboration on the molecular pathology of BSE . . . should be allowed to get underway before offering material to outsiders. I know that healthy competition may be a plausible reason for giving material to other workers, but past experience suggests that collaboration with Prusiner degenerates into unhealthy acrimony." Continuing with the policy of refusing to cooperate with outsiders that began the earlier exchange with Marsh, scientists in the United Kingdom may have begun to wonder if another Nobel Prize, like Gajdusek's, might not be available for studying this disease. But the torrent of requests for material kept coming, and eventually the ministry had to relent. By the end of 1988, the ministry had changed its policy and had begun to send material out to other researchers.

Although various drafts of the Southwood report had been circulating around the ministry throughout October, the committee did not meet formally until the following month. They tried to summarize "what we know so far" and they tried to project into the future—always a dangerous game when dealing with an unknown epidemic. "A constant number of cases, of the order of 350–400 per month can be expected; this is an incidence of one case per 1000 adult cows per year," the report noted reassuringly. "This rate of presentation of the disease will continue until 1993, a cumulative total of about 17,000–20,000 cases from cows currently alive and subclinically infected. Thereafter, if cattle-to-cattle transmission does not occur, then a reduction in incidence would follow with a very low incidence in 1996 and the subsequent disappearance of the disease."

In hindsight, the predictions of the Southwood report are laughable underestimates, but in 1988, figures of 17,000–20,000 were alarming to the ministry. After all, their continuing public posture was that BSE was scrapie and really nothing to worry about. The last thing the ministry wanted was for these kinds of "huge" figures to be released to a complacent and unaware public. The British public was continuing to enjoy the meat pies that they loved and the traditional English roast beef was still being avidly consumed. The ministry had no intention of rocking that boat by releasing such alarming figures to the public. The Southwood report also noted reassuringly that the "risk of transmission to humans (of BSE) appears remote." This was another phrase that would come back to haunt them.

As the rendering industry began to feel the bite of the restrictions on their meat and bonemeal product, sales began to drop. So they, in turn, began to exert pressure on the ministry. In a November 1988 letter, the industry tried to convince the ministry to lift the ban imposed on them, citing the "blight" all the adverse publicity had caused on their industry. They also complained that they were being forced to export their products overseas at uneconomical prices because of the widespread suspicion in other countries that their product was contaminated with scrapie or BSE. It appeared that the rendering industry's main concern was with the dropping prices for meat and bonemeal product. In an attempt at negotiating, they offered to voluntarily remove spinal and brain tissues from all their meat and bonemeal product, reminding the minister of the temporary nature of the ban and also reminding him of the ministry's position that the majority of meat was safe, once the brains and spinal cords had been removed.

As the BSE cases mounted, the disposal problem escalated. After the idea of using lime pits for the disposal of carcasses was abandoned, the ministry retreated to simply burning the carcasses as quietly and as unobtrusively as possible. In remote places, the carcasses were beheaded and the heads sent separately to vet pathology facilities. The rest of the carcasses were burned in the open. As the number of carcasses mounted, however, and the wind began to blow the overpowering stench in the direction of populated areas, the public began to complain loudly about the stink. The trucks moving the carcasses to the funeral pyres were dropping copious amounts of blood, bodily fluids, and the occasional body part onto the roads. This disgusting and smelly gore provoked an intense public reaction, especially in Cornwall, Devon, and the southern counties in England. They were the counties experiencing the most BSE-positive animals and hence the most late-night trucking of headless carcasses.

Eventually the ministry began to formulate a policy of building concrete incinerators to burn the carcasses. "The maximum number of carcasses transported in one load so far is 28," according to an internal ministry memo discussing the logistical problems of transporting carcasses in the southwest. "The only problems encountered are narrow access roads as the collection site is approached. This was exacerbated by a local farmer effectively preventing access from the easier direction by constructing a massive concrete wall on the inside of a right angle bend. . . ."

Photographer David Jackson brought out the full horror of the secret ministry funeral pyres when he published the picture of the headless corpses with their legs sticking up in the air. When it appeared on TV and in newspapers at the end of 1988, the photograph ushered in to the public's mind the full horror of the BSE epidemic. The photograph became an emblem for the secret operation conducted by the ministry and the surreal scene of smoke drifting over the burning cattle legs provoked public outrage. The photo marked the end of the open funeral pyres and the beginning of the incineration of the carcasses. Jackson's photo did much to wake up the British public to the epidemic that was around them. However, as the number of cases escalated, the incinerators became stretched to the breaking point and the ministry began to dig large open graves to accommodate the overflow. This policy provoked concern. By early 1989, the ministry was disposing of about 130 carcasses per week and they projected the number would reach more than 200 by 1990.

As the ministry continued to dispose of the affected carcasses, press reports routinely speculated about the dangers of transferring the disease to humans. These reports were followed by standard ministry assurances that there was a negligible chance of transfer to humans. The number of BSE deaths was escalating every month and a December 1989 article in *British Farmer* complained: "Many farmers felt they could accept the loss from one BSE case. Now that some are losing several cattle in a matter of weeks, BSE is causing serious financial problems. . . . Recent Ministry figures revealed that BSE had been confirmed in 6381 cattle from 4301 farms."

The staggering escalation of the epidemic in 1989 had completely belied the assurances of the Southwood report. In just twelve months, almost half their projected total of 17,000–20,000 cases had been reached and still the epidemic showed no signs of stopping. January 1989 marked the announcement of the first case of BSE in the Republic of Ireland. The animal, a four-year-old Friesian, died within a couple of weeks of the diagnosis and pathology studies confirmed BSE. At that time, five cases had already been confirmed in Northern Ireland.

Throughout 1989 and into 1990, the flurry of internal

memos, often several per day circulating in the corridors of the ministry, belied the sporadic TV appearances of the chief veterinary officer who repeated the mantra that BSE was essentially scrapie, that it does not transmit to humans, and therefore there is nothing to worry about. The internal panic and the calm TV pronouncements were a study in contrasts. The rendering and pharmaceutical industries resisted sudden mandatory restrictions on their practices. A ministry mandate in 1989 that prevented any kind of offal (brain, internal organs, etc.) from entering the British food supply was issued but only halfheartedly enforced. The mandate came without precise guidelines.

There was, in fact, no way of enforcing this ruling. Cows going to the slaughterhouse still had heads split open with meat cleavers and chain saws in an attempt to remove the brains prior to processing meat for human consumption. There were comic-book scenes of slaughterhouse workers trying to use the same hole they had used to kill the animal, by shooting a bolt through the skull, as a means of sucking the brain material out, prior to sending the hapless animal farther down the assembly line to be processed into meat for human consumption.

These crude attempts at meeting the ministry directives resulted in brain material being splattered on the outside of the cow skulls. The meat from these skulls was then used in the manufacture of hamburgers and other meat for human consumption. The practical reality was that the ministry ban on offal combined with the low-tech methods of achieving the ban actually increased the likelihood that hamburger meat would be contaminated with brain residue.

And then there was the question of compensation. Would the dairy farmers be compensated for destroying their investment? The ministry directive back in 1988 that farmers would only be paid 50 percent of the market value of the animal actually encouraged people to slip infected animals through the loopholes. Farmers were already hurting from the BSE epidemic. Were they going to obey the ministry guidelines and lose half the value of their animals by reporting BSE-positive cases? Certainly not. Strategically, critics believed that the 50 percent compensation package doomed the program to failure from the start and resulted in more BSE-positive animals being shipped to slaughterhouses.

Meanwhile, Dr. John Wilesmith was trying to get an ambitious study off the ground to examine whether the BSE agent was being passed from infected mother to offspring. But the ministry was starved for funds. Meeting after meeting was called and dozens of memos went back and forth. Still the study was not funded.

In May 1989, *The Times* reported that "Meat from diseased cattle may be on sale to the public," and warned about the increasing outrage felt by health professionals in Britain toward the ministry's lax response to the BSE crisis. "Senior pathologist and veterinary surgeons yesterday accused the government of not doing enough to stop cattle which are infected with an incurable brain disease from being processed for human consumption. Some said that they had changed their own eating habits because of the health risk." The article went on to repeat Sir Richard Southwood's statement that the chances of BSE passing to humans were "remote." The article also criticized the government for allowing the export of meat and bonemeal to other countries, in spite of the Southwood committee specifically pointing to meat and bonemeal as the source of the BSE epidemic in Britain. This was one of many loopholes that would later come back to haunt the ministry.

In mid-1989, several newspaper articles began to point to the dangers of cow brains ending up in sausages and meat pies. *The Guardian* newspaper even leaked a story claiming that the ministry was getting ready to ban cow brains in meat pies. The ministry denied the allegation in an internal memo that was never made public. The sale of meat pies and sausages continued and these meats were eaten as if nothing was happening in the British countryside. Meanwhile, the disposal of carcasses was becoming so onerous that the ministry appealed to the British Ministry of Defense (MoD) for help in the matter. The MoD promised to cooperate.

The Times was one of the first newspapers to use the phrase "mad cow" in a headline. Other newspapers had begun to use the phrase sporadically in the body of their articles. *The Times* exposé proclaimed: "A ban on the sale for human consumption of cattle brains and other organs believed to carry the 'mad cow' disease, bovine spongiform encephalopathy, is expected to be announced soon by John McGregor." The article went on to note that "four months ago the government announced a ban as a cautionary measure on the use of bovine brain, thymus gland, spleen or spinal cord tissue in baby food." In mid-1989 it was questionable how many people in Britain even dreamt of the possibility that baby food could contain any bovine brain, thymus, spleen, or spinal cord. But the world of the ministry in the late 1980s became increasingly surreal by the month. Anything seemed possible. After June 1989 headline writers used the phrase "mad cow" almost continuously. It was certainly more evocative than the staid term "BSE."

Gerald Wells, who had first diagnosed BSE back in 1986, spent a couple of weeks in the spring of 1989 in the United States on a quiet trip to assess the likelihood of BSE occurring in the United States. His confidential thirty-three-page report makes interesting reading. Wells toured many of the USDA scrapie research facilities, visited Marsh in Wisconsin, and learned about Marsh's ideas on a presymptomatic form of BSE in the United States. Wells also visited the scene of the Stetsonville mink outbreak and learned that multiple animal-to-animal transmission experiments were ongoing in several parts of the United States. Wells discovered that several intense mink-to-cattle transmission experiments were taking place in a remote USDA outstation at Pullman in Washington State.

Lastly, Wells visited two wildlife research facilities in Colorado and Wyoming, separated by about 150 miles, at which a mysterious new spongiform disease had developed in captive deer. The disease was named chronic wasting disease, or CWD. The Wells report was thorough and because of the sensitive nature of the findings, many of which had not yet been published, was kept on extremely close hold at the ministry.

In July 1989, the ministry became alarmed at an urgent telex from the European Agriculture Commission that suggested action needed to be taken regarding the wholesale export of meat and bonemeal product from Britain to Europe following the banning of the product in Britain. Ministry officials objected to the "tone" of the telex because it appeared to be concerned that Britain was exporting BSE to Europe via the very same meat and bonemeal product that had been banned in Britain.

"The rendering industry has survived the July 1988 prohibition," the ministry memo continued, "in part because they have been able to fill the gap in the market through exports [to Europe]." Not only was the ministry not blocking the export of meat and bonemeal potentially contaminated with BSE to the European countries, but they actually seemed to be encouraging it. The reason for their consternation at the urgent tone of the telex from Europe was obvious: If they stopped exporting potentially BSE-infected meat and bonemeal to Europe, then the profits of the British rendering industry might be adversely affected. This appeared to be another example of the surreal atmosphere that lay behind the closed doors of the British Ministry of Agriculture. At the end of July, the Europeans followed up on their telex by banning the importation of cattle from the United Kingdom and by insisting that the only cattle that could enter Europe were from certified BSE-free herds.

By early 1990, as the number of BSE cases climbed, the British press had begun claiming that the government was misleading the public. Several press articles in January 1990 accused the government of ignoring unscrupulous farmers who were deliberately selling diseased animals to the slaughterhouse knowing they were infected. Headlines like "Mad Cattle Meat Racket" and "Diseased Cattle Sold" screamed at the hapless consumers as they headed for the supermarkets to buy their meat. An article in the January 4, 1990, issue of *Today* reported: "Greedy farmers are selling beef contaminated with BSE, the mad cow disease, for human consumption, it was revealed today." The media drumbeat, amazingly silent since 1985, began to intensify. The Ministry of Health decided to launch an inquiry into the possible threat to human health from mad cows, as more than 500 animals were dying every month. This was an embarrassment to the Ministry of Agriculture, which days earlier had trotted out its routine assurance that the risk of passage to humans was "remote." The Ministry of Agriculture was coming under severe pressure to review its policy of only giving 50 percent of the market costs of beef to farmers. It was this policy, of course, that many believed was inciting farmers to try to bypass the ministry rules and sell their diseased animals for public consumption. In January 1990, David McLean, the parliamentary secretary for the ministry, rejected any suggestion that the 50 percent compensation would be increased. "We believe that fifty percent of the value of the animal as if it were healthy is fair compensation for an animal that is terminally ill and therefore worthless," McLean declared.

Another blow to the ministry's carefully orchestrated fiction came when the United States banned their military personnel from eating British beef. The order that all hamburgers and steaks would henceforth be flown in from the United States angered beleaguered British farmers, many of whom depended on the $3-million-a-year market.

When the press reported that experiments had successfully transmitted BSE from cow to cow through cerebral injection, the ministry's response was to downplay the importance of this study by citing the "unnatural" mode of transmission and maintaining that the risk of transmission to humans was "remote." More information began to emerge in Britain that cosmetics, particularly antiaging creams, contained 0.1 to 5 percent extracts of bovine spleen or thymus. The cosmetic industry had kept this kind of information low profile, but increasing press focus forced it out into the open.

In February 1990, the newspapers began trumpeting a new study that showed mad cow disease could be transmitted to mice by their eating cow brains. This study provoked a fresh round of speculation in the media that humans were next. This, in turn, forced a fresh round of denials from the ministry. Again they cited Sir Richard Southwood's by now famous statement that the risk of transmission to humans was "remote." Several investigative TV shows began to appear for the first time, each one ratcheting up the pressure on the ministry. One particularly hard-hitting segment of the *World in Action* TV show in January 1990 provoked a media blitz saying in effect that it was miraculous that British people were still eating beef, given the obvious risks.

In midmonth, the then agriculture minister John Gummer finally came to his senses and doubled (to 100 percent of current market price) the compensation for farmers who reported BSE-infected animals. This removed the incentive that had been such a problem for the previous two years and that was probably responsible for significantly accelerating the epidemic.

In March 1990, five exotic animals died in three zoos in southern England. A gemsbok, an oryx, a kudu, a nyala, and an eland all died within a few months of one another after eating meat and bonemeal. The media emphasized the "species jump" to these animals and once again raised the possibility of a species jump to humans. And to make matters worse, a couple of months later, in May 1990, the newspaper headlines screamed "Mad Cat Disease." "The government moved to reassure the owners of seven million cats yesterday," the *Financial Times* story stated, "when it emerged that a five year old Siamese had died after developing the feline equivalent of mad cow disease." The culprit, it emerged a few months later, was pet food. The specter of the disease striking down the family cat brought the possibility of a jump to humans one step closer to the British public.

Within a few weeks, according to a story in *The Guardian*, the National Health Service announced that certain schools in Britain would begin to remove beef from the school cafeteria menus. This announcement provoked more outrage from the beleaguered farmers as well as protests from the ministry. It seemed every week brought another nail in the coffin of the ministry's credibility. After all, the ministry had been saying the risk to humans was "remote" for a couple of years now and here the British school system was plainly skeptical of their public pronouncements.

With the almost constant clamor in the press of the various species jumps of BSE, it was almost inevitable that sales of British beef would begin to fall. By the end of May 1990, industry sources calculated they had lost about $100 million and this was probably just the beginning. Agriculture minister John Gummer's frequent appearances on TV to reassure the public that British beef was "safe to eat" were not ringing true anymore. The National Health Service's chief medical officer, Sir Donald Acheson, issued a statement meant to shore up flagging British morale: "I have taken advice from the leading scientific and medical professionals in this field. I have checked with them again today. They have consistently advised me in the past that there is no scientific justification for not eating British beef and this continues to be their advice. I therefore have no hesitation in saying that beef can be safely eaten by everyone, both adults and children including patients in hospital." Sir Donald's ringing endorsement of British beef carried a note of desperation as the British public began to see the light five years after Jonquil's death.

In June 1990 the French joined the Germans in banning British beef and others in the European community were threatening to follow suit. The value of the French market alone was approximately $350 million, so the French move drew howls of protest from the British farmers. As the months wore on, a sense of inevitability was setting in.

At the end of the month, Professor Richard Lacey at the University of Leeds, who had long been a thorn in the side of the ministry, gave testimony at a parliamentary inquiry into BSE. Lacey challenged the Southwood working group's ability to assess risks for humans and challenged the central assertion of the group that "BSE came from scrapie." Lacey asked why the Southwood committee was not reconvened to determine why they were so woefully off the mark in their predictions of the number of BSE cases. Lacey further challenged the Southwood committee to determine why they had engaged the British people in a long-term epidemiological experiment with BSE. His anger that the British people were essentially being used as guinea pigs would resonate even more strongly in a few short years.

At the same time, a group led by Dr. John Collinge published an article in *The Lancet* suggesting that current methods of testing for CJD, the human form of BSE, were probably leading to an underestimation of the numbers. Collinge's paper stated that there were, in fact, far more people dying of CJD in Britain than the available numbers suggested. "This observation may be relevant to the assessment of possible transmission of bovine spongiform encephalopathy to man," Collinge concluded with typical British understatement.

G. R. Roberts of St. Mary's Hospital in London responded to Collinge's article with a letter in *The Lancet*, also claiming that CJD was widely underestimated in Britain. The ministry hotly disputed Roberts's findings. They claimed that Roberts was not a medical doctor (merely a Ph.D. researcher) and, secondly, that this had nothing to do with BSE anyway since BSE cannot be passed to humans. This response was an exercise of circular nonlogic that had come to characterize the ministry's increasingly defensive mentality.

A couple of months later, Gerald Wells, the pathologist who had first found the spongy neurons in the BSE-infected cows, announced that BSE had been successfully transmitted to pigs via brain injection. This finding, in turn, led to another media barrage about "mad pig" cases in the United Kingdom. At the end of September 1990, ministry officials had sent BSE material from infected cows over to Gibbs and Gajdusek at NIH so they could run it through the chimpanzee and monkey tests they had become so adept at. The ministry lamented it could be years before the results were known.

On November 14, 1990, the ministry received word of a single positive BSE case in Switzerland and a further fourteen that were likely positive. The disease had spread beyond the British Isles and was now in mainland Europe.

Meanwhile, Dr. Richard Lacey continued to hammer away at the ministry and at their incompetence in continuing to allow an *always* fatal disease into the food chain of unsuspecting British people. The ministry was oblivious, responding that Dr. Lacey's latest paper and public attack had contributed "nothing new" to the debate. In essence they continued to ignore Lacey as well as the increasing chorus of concerned physicians and scientists who supported the position that the ministry should be more proactive.

As the ministry continued to conduct internal epidemiological studies, France confirmed two positive BSE cases in Normandy on March 4, 1991. The ministry, however, were more concerned about recent reports that had come from Gajdusek's group in the United States, showing that the soil remained infectious even three years after burial of hamsters infected with scrapie. Concerns were beginning to spread about the transmission of BSE in the soil from the numerous open graves that farmers were digging.

Routinely, public outrage would filter to the surface after the ministry was caught secretly trying to burn or bury large numbers of BSE-infected carcasses. The cat-and-mouse game of the ministry not wishing to appear concerned while at the same time trying to bury the hundreds of carcasses would be comical if it were not so tragic. At the end of March 1991, data began to trickle in apparently confirming that BSE was transmissible vertically from mother to offspring. The ministry issued a press release emphasizing the few cases they had found so far. In the release they gave the figure at 25,826 confirmed cases of BSE as of March 18, 1991, with only a few dozen where vertical transmission had occurred. The huge figure of more than 25,000 confirmed already exceeded the Southwood projection for the entire run of the disease.

The ministry was now grappling with two major problems. The first was the increasing number of BSE cows that were born after the feed ban of July 18, 1988, was put in place. These animals were a severe embarrassment to the ministry because the publicity overseas was directed at the number of loopholes in the feed ban. The BSE-positive cases were a graphic and painful reminder that the imposition of the feed ban had not solved the problem. Other countries were using these new cases to justify an extension of their bans of importation of British meat and animals. This in turn was putting a lot of pressure on the ministry. But the figures could not be denied or hidden. Domestically, the new BSE-positive cases provided strong ammunition to the ministry's increasingly numerous critics. The new statistics were a reminder to the public that the feed ban had a large number of loopholes and was basically unenforceable.

The second major problem the ministry faced was the rendering industry's shoddy safeguards. The equipment and technology used by the British rendering industry in late 1991 failed many of the basic tests required by European Union regulations. The low-temperature treatment of meat that was routine in British rendering plants would fail the more stringent European temperature guidelines. The ministry worried that this could have disastrous consequences on the British rendering industry's ability to export any of their products. The ministry realized that only a huge capital influx would solve the problem. In short, the entire British rendering industry was at risk.

The ministry's protestations that British beef was perfectly safe fell on deaf ears overseas. In January 1992, the European Union put a large package of food aid together for the Russian public. The aid package included about 2,000 tons of British beef. The Russians promptly blocked the arrival of the British part of the beef consignment on health grounds. "The consignment of British beef rejected by the Moscow authorities on health grounds was yesterday unloaded 600 miles away in the Arctic city of Murmansk," reported *The Times*. "All shipments of food aid from Britain meanwhile have been suspended." This rude rejection of British beef by Russians who were starving was particularly galling to the ministry. They sent Keith Meldrum, their top bureaucrat, over to Moscow to try to diffuse the Russians' concern. To rub salt into the wound, a telex from Riyadh, Saudi Arabia, also announced a ban on the importation of British beef. As the international embarrassments piled up, John Wilesmith's epidemiology division at the ministry circulated a series of confidential memos that predicted more than 30,000 new cases of BSE per year for the next couple of years. The epidemic was showing no signs of slowing.

In February 1992 a confidential memo circulated around the ministry reporting the first successful transmission of BSE to primates. This was a potentially troublesome finding for the ministry to explain to an already panicked British public. Most of the internal discussion involved controlling the dissemination of the information to the press to limit the damage.

Germany confirmed its first case of BSE in April 1992. The epidemic was slowly but surely gaining a stranglehold on mainland Europe. France, Germany, and Switzerland were now coping with the spread of the deadly contagion through their cattle industries. Although there were no public recriminations as yet, it was a safe bet that these European countries would eventually hold the export of contaminated meat and bonemeal from Britain, after it was banned in Britain, at least partly responsible for the devastation of their dairy industries. The decision by the British renderers to minimize their losses, aided and abetted by the Ministry of Agriculture, would reap consequences down the road.

That same month Dr. Richard Lacey launched another angry diatribe against the ministry, accusing the department of "complacency" and of allowing a deadly agent into the food chain, including children's food, in Great Britain. Publicly, the ministry ignored Lacey's attack, but privately they were outraged. "Professor Lacey's speech is based on no new information; his assertions remain as questionable as they always have been," stated a memo from the Animal Health Disease Control Division. "Professor Lacey makes accusations of complacency without acknowledging the action that has been taken."

In June 1992, the ministry released the latest figures on BSE incidence. There were now 631 new cases per week being reported. The British public was becoming increasingly uneasy. A press release from the ministry urged the public to keep calm over these new figures and to continue buying British beef. "Shoppers were today urged not to panic over a new mad cow alert," reported Britain's *Evening Standard*. "They were told to carry on buying and eating beef—20,000 tonnes of it a week—despite claims that the number of new cases had risen to an average of 631 per week. . . . A Ministry spokesman today assured the public: 'beef is perfectly safe to eat in this country. It is safe for adults, for children and even those in hospital.'"

In September 1992, Denmark reported its first confirmed case of BSE. The pedigreed highland cow had been imported from Great Britain in June 1988. A few months later, Mexico announced a ban of all milk and milk products from Britain. The international embargo was now a steamroller.

The end of 1992 brought little solace to the ministry. The past four years had resulted in unprecedented spread of BSE, with almost 100,000 new cases reported. The epidemic had crossed the sea, first to Ireland and then to several countries on mainland Europe, including France, Germany, and Denmark. The years had seen a severe erosion of the public's trust in their government as country after country ignored the ministry's protestations and began banning cattle, beef, milk products, and feed from Britain.

The storm clouds that had gathered over Britain regarding the BSE epidemic were now about to break.

13

The Tipping Point

Vicky Rimmer was blonde, vivacious, and only fifteen years old. She had her whole life to look forward to. She loved shopping with her friends and she much preferred rock music to homework. In May 1993 Vicky developed a bad cough and then sharp neck pains. The local doctor could find nothing wrong with her so he advised her grandmother, Beryl Rimmer, to send her back to school. The symptoms started slowly and then accelerated over the summer. She started forgetting and couldn't remember where she had put things. She had increasingly long dizzy spells. And her personality seemed to alter.

Vicky started losing weight over the summer, and after nearly falling off a playground ride in early August 1993, she was hospitalized in Wrexham, Wales. By the end of the month she was down to eighty-four pounds and had great difficulty swallowing. Finally, a government doctor diagnosed her condition: fatal spongiform encephalopathy. Beryl Rimmer had never heard of it. Upon returning from the hospital, Vicky's local doctor translated the diagnosis into more understandable language: It's "mad cow disease." Creutzfeldt-Jakob disease was practically unheard of in one so young. By the end of 1993 Vicky had slipped into a coma. "I am convinced she got the disease from a burger," Beryl Rimmer told the *Daily Mail*. "She's never been ill in her life. She was the healthiest child on Earth. But she's always loved meat."

By 1993, BSE had already killed about 120,000 cows in Great Britain, but it was Vicky Rimmer's dramatic illness that stopped the public short. As Vicky's story made the news, word began to leak out that three dairy farmers, some of whom had BSE-infected animals on their farms, had contracted and died of CJD. Then in May 1995, the *Today* newspaper in Britain reported on another case of CJD: "Scientists are investigating the death of a teenager who showed possible symptoms of a human form of Mad Cow Disease. The 19-year old student died in hospital on the weekend. The student is the second British teenager to suffer symptoms of Creutzfeldt Jakob Disease within 18 months." Vicki Rimmer was still in a coma.

These isolated cases of CJD had not yet alarmed the Ministry of Agriculture, Fisheries and Food, which was more preoccupied with damage control after its discovery in June 1995 of BSE in a young cow that was born in 1992. This was doubly embarrassing to the ministry since they had loudly trumpeted to the British public their certainty that no more BSE cases would emerge after 1988. But even more worrisome, they had gone to great political lengths to persuade the German government to accept export of live cattle born after January 1992 on the (faulty) premise that these cows were now BSE free. This latest case of CJD was the subject of a group of memos exchanged in the ministry, which eventually decided not to publicly announce the case. It was becoming obvious that there were just too many loopholes to stop the epidemic.

A memo that circulated through the ministry in July 1995 announced that *the cumulative prion load in an animal outside the brain was probably about equal to that in the brain.* That meant that the reliance on just removing brain and spinal cord from the animal before it entered the food chain was not sufficient. High prion loads had been detected in gut-associated lymph tissue in BSE-infected cows. This had major implications since the British had a long-standing tradition of eating "tripe." Tripe is the lining of a cow's stomach and thousands of Britons relished their meals of tripe and onions every day.

In an effort to find out how serious the problem was, the ministry launched a nationwide survey of slaughterhouse practices. The inspectors visited the premises of 392 slaughtering facilities and observed how the workers were separating the intestines and offal from the meat. What they found was disturbing. Ministry-mandated procedures were not being followed in 309 out of the 392 plants they visited. "The overall impression of this snapshot view of the industry," the ministry's memo concluded, "is that there is widespread and flagrant infringement of the regulations . . . there are grounds for suspecting that the highest risk tissues (brain and spinal cord) have been mixed with other by-products and processed for animal consumption." Whether the industry was deliberately dragging their feet in implementing the ministry directives was unknown, but the survey showed the disastrous state of the attempts to stop BSE-infected tissue from circulating back into animal feed.

Then in August 1995, *The Times* announced that "A woman who used to prepare meat pies is thought to be suffering from Creutzfeldt-Jakob disease, the human form of Mad Cow Disease." This article appeared within a few days of a documentary TV program by the prestigious *World in Action* program titled "CJD in Teenagers." The combination of newspaper and TV coverage about human mad cow disease set off alarm bells in the public.

The following month the ministry discovered a fourth case of CJD, this time in a fifty-nine-year-old farmer in Gwynedd, Wales. Like the other three cases, this latest one also had had a BSE-positive case in his herd, but notably, the man was a cattleman, not a dairy farmer. Memos circulated through the ministry noting that this fourth case was likely to increase the publicity since a *Lancet* article was due out the same week that described the first three cases of CJD.

Within a few weeks, multiple newspaper articles appeared about the increases in "human mad cow" diseases detected in the British population. Headlines blared that mad cow disease in humans was at a new high. These stories were in reference to an annual survey of CJD cases that showed a modest increase in sporadic CJD cases in Britain. The figures were probably meaningless, since they likely vastly underesti-mated the true figures, but the press amplified the level of concern. By the end of October 1995, several newspaper arti-cles announced the fourth case of CJD in a farmer and the resulting public pandemonium forced the ministry to release a terse news flash: "CJD is a disease in humans. The incidence of CJD in dairy farmers is no higher in Britain than in other European countries. BSE is an animal disease." The short, clipped tone of the press release seemed to underline the min-istry's defensiveness. There were no conciliatory statements to the public about how understandable the public concern was. It appeared the ministry was once again under siege.

On November 10, a curt memo circulated through the ministry regarding a newspaper article: "The *Today* newspa-per today reported two suspected CJD cases: a 30 year old male who has recently died in Belfast and a 42 year old male in Liverpool who we understand is currently in a nursing home. Our usual line in such cases is appropriate: 'We are aware of these cases. The necessary investigations are under-way and we cannot comment further at this stage.'"

There seemed to be a note of resignation in the business-as-usual reaction to two more CJD cases in relatively young people. The ministry was fighting a losing battle against time. And interestingly, the British press was now com-pletely driving the reporting of new cases. Each new case of human mad cow disease was brought to light first in the press, followed by a bland denial from the ministry. This left the public with an awful doom-laden feeling, knowing that they could probably not trust the government but hoping that it was telling the truth in any case.

In response to the latest CJD cases, Dr. Richard Lacey, the longtime thorn in the side of the ministry who had been warning for years about the dangers of eating contaminated beef, penned a letter for *Today*. "More than five years ago I demanded that the control of BSE and its frightening potential to decimate the human population was through slaughtering of all infected cattle herds," Lacey began. He explained that action was not taken because it was deemed too expensive and the ministry was too preoccupied with protecting the cattle industry. "Instead," Lacey continued, "the ministries have orchestrated a campaign of deception, misinformation and manipulation of cowardly scientists. . . . We still live in a class ridden society. The top class is the government and the meat industry. The bottom, or experimental class is the already infected consumer (i.e., most of us) waiting for the terminal dementia beginning as early as the teens with no diagnostic test, no vaccine and no treatment." Lacey's words were especially powerful because for five years he had very publicly castigated the ministry for exactly the nightmare scenario that was beginning to unfold.

In a remarkable press release dated November 19, 1995, the British Ministry of Health completely rejected any suggestion of a link between BSE and human CJD: "Beef is beef is beef. Be it prime cuts, mince, burgers, pies or sausages. All play a key role in providing the nation with a healthy balanced diet. Enjoy it. Just as you have always done." The apparent dismissal of the mounting data that suggested there was in fact a link between BSE and CJD is breathtaking.

Three weeks later, Prime Minister John Major stood up in the House of Commons and strongly rejected any link between BSE and CJD. By that time the mad cow scare was occupying newspaper headlines every day, and the government began to fear that this was not going to go away. Even John Major's strong statement in the Commons was not sufficient to quell the furor.

The death of another thirty-year-old from CJD on February 15, 1996, this time from Belfast, brought to four the number of recent cases of CJD in people thirty or younger, including teenagers. A ministry memo describing the latest case warned that the parents were very upset and were preparing to mount a publicity campaign. "In the meantime," the memo concludes, "there is not much we can say, other than the case is being fully investigated and offering sympathy to the family."

Then the bombshell hit. The memo, dated March 8, 1996, was written by Dr. Eileen Rubery, a highly experienced policy maker who had advised the ministers in Whitehall for more than fifteen years, following a meeting of the Spongiform Encephalopathy Advisory Committee, which had ended three days earlier. The memo described in chilling detail the identification of a new (amyloid-plaque rich) type of CJD disease (clinically and pathologically) occurring in under-forty-year-olds over the last two years. "This could be called atypical CJD (A-CJD)," stated Rubery. This "atypical" CJD appeared to have much more "florid" plaques in the brain, as well as evidence of neural degeneration in the spinal cord. Finally, the patients were much younger than the traditional "classical" CJD patients and they took longer to die. This "atypical" CJD would later be renamed nvCJD (new variant) or simply vCJD.

Rubery's memo soberly stated that if the two new cases that they were working on were confirmed, that would raise to twelve the cases of atypical CJD in people under forty that the group had so far encountered. "The identification of this new type of CJD," Rubery continued, "makes a link with the BSE epidemic a likely hypothesis although it is not the only hypothesis. The public will see this as proving the link. It is probably prudent that our policy is now developed as if that link were present." In the well-written memo, Rubery elaborated on her hypothesis: "Atypical CJD is caused by the BSE agent jumping species barrier from animal products to humans via food or a parenteral (intravenous) route. The ten cases so far are the beginning of an epidemic in man."

At the end of the memo, Rubery, who was writing up a scientific paper on the data for submission to *The Lancet*, discussed the timing of release of this bombshell to the public. On public health grounds, as well as political and credibility grounds, Rubery asked if the minister could wait the four to five weeks it would take to get the paper published "before informing the public." The answer would be "no." But it is still unclear why this extremely abrupt U-turn by the Spongiform Encephalopathy Advisory Committee took place in March 1996, since most of the young people on whom their conclusions were based had died in 1995, had been autopsied at the time, and were heavily reported by the media. Somebody had been accumulating the data on their unique pathology, how these new cases differed from "sporadic" CJD. Could it be simply that someone sat on the explosive data and delayed announcing the link between eating beef and CJD?

In any case, the tone of Rubery's memo differs from all

other memos that circulated from the government in that it seemed to finally accept the reality of a very urgent problem. It frankly discusses the possibility of the beginning of a new epidemic in humans and questions the potential size of the epidemic. It is notable that the copy of the memo that was eventually released to the public has numerous handwritten notations in the margin by an unnamed bureaucrat that constantly question Rubery's use of the word *epidemic*.

Rubery's memo about the Spongiform Encephalopathy Advisory Committee meeting marked a turning point in the long history of BSE in the United Kingdom. From April 1985, when Jonquil's symptoms first became noticeable, until March 1996, almost eleven years had passed during which an attitude of patrician and condescending denial had pervaded both the Agriculture and Health ministries. The Rubery memo came just four months after the press release quoting the arrogant and self-serving statement from the chief medical officer, which asserted unequivocally "British Beef Is Perfectly Safe to Eat." Now finally, in March 1996, the ministry perceived a problem. Rubery's memo appeared to unfold all of the dire predictions that Richard Lacey had made over five years, and Lacey had been dismissed and categorized as a crank by the ministries of both Agriculture and Health.

The alarming news then began to ascend the chain of command. On March 18, 1996, in a letter to the prime minister, ministers Hogg and Dorrell began: "You should be aware of a very serious development on BSE. In brief, the CJD unit had identified a new variant of CJD in young people in the UK. . . ." Two days later, Health Minister Stephen

Dorrell stood before an astonished House of Commons and delivered the bad news to the British people. The statement in Parliament was accompanied by a flurry of press releases from both health and agriculture departments.

"We Have Already Eaten 1,000,000 Mad Cows," blared the *Daily Mirror* headline that day. The announcement triggered a widespread panic and the bottom fell out of the British meat market. The public reaction was rapid and visceral. "Why weren't we told there might be a link?" asked Ronny Richardson, whose wife, Anne, had succumbed to vCJD. "If the government had been upfront we would have stopped eating beef. They kept it under wraps to protect the beef industry. I am livid. I feel like I have been misinformed and patronized and both my wife and I have been put through a lot of unnecessary suffering as a result. We meant nothing to the government. It is a disgrace."

Another CJD victim's mother, Nora Greenhalgh, publicly and explosively displayed a letter she had received from John Major's secretary denying any link between beef and CJD. "I didn't believe him then," she said, according to authors Rampton and Stauber (*Mad Cow U.S.A.*). "I don't believe him now. This was a deliberate cover-up to avoid tarnishing the reputation of British beef."

Recriminations flew, stretching back to the ministry's original insistence that eating beef was safe and that the BSE epidemic would never reach more than 20,000 heads. The accusations prompted a defensive letter from Sir Richard Southwood in the *Daily Telegraph* on March 28, 1996, as he tried to escape the firestorm of outrage that was pouring in his direction. "May I set the record straight about my role as chairman of the 'mad cow disease' working party?" his letter asked. The plaintive tone of Southwood's defensive letter seemed to show Southwood had missed the point.

So great was the rage directed at the government's long history of dissembling on the problem that Health Minister Stephen Dorrell complained: "It isn't the cows who are mad, it's people who are going mad, what all of us have to do is step back from the hysteria and believe the facts." Again the arrogance and patronizing attitude of the bureaucrats shines through loud and clear.

As Vicky Rimmer still lay in her coma fighting for her life in the hospital, Beryl Rimmer responded to Stephen Dorrell's complaint. "Dorrell is such a swine," said Beryl, according to authors Rampton and Stauber. "He can't have any feelings. I only wish he could see my Vicky lying helpless in hospital. One look at her would change his mind."

Later it emerged that the government had tried to persuade some parents of the CJD victims, Beryl Rimmer included, to keep quiet about the disease. Pulitzer Prize–winning journalist Richard Rhodes has reported that the physician investigator who visited Beryl Rimmer and who diagnosed spongiform encephalopathy in the young comatose Vicky Rimmer had pulled Beryl aside and told her to keep quiet about it. "Think about the economy," he had told Beryl Rimmer. At some level, probably for years, the government had known there was an impending disaster.

Vicky Rimmer died in November 1997. By 2004, she was one of about 150 people who had died from variant CJD (vCJD) in the United Kingdom and Europe. (In Switzerland, where the number of mad cow cases has been reported honestly by the government, the number of sporadic CJD cases has doubled.) The projected number is unknown. In the past few years there have been a number of gleeful predictions that the annual death rate from vCJD was dropping and therefore the problem was not that great after all. But the latest figures from early 2004 show that the downward trend has reversed.

On an even more disquieting note, one has only to look at the average incubation period of a couple of decades for kuru in Papua New Guinea to see that we may only be seeing the tip of the iceberg. In spite of the cessation of cannibalism in the 1960s, there were still documented instances of kuru death thirty years later. If the majority of people began to be infected by eating contaminated meat in the late 1970s or early 1980s, then it is still too early to predict when the peak of CJD cases will be reached, though it's likely to occur by the end of this decade. And of course, the increasing number of calves infected with BSE born in the United Kingdom *after* 1988 showed that the restrictions imposed by the agricultural ministry were not nearly stringent enough. In fact, multiple calves born in Britain in 1993 were subsequently shown to have BSE. Therefore, the current official line that the vCJD epidemic is already on the decline in the United Kingdom is not convincing.

But that's not the end of the story. There is also a growing suspicion among researchers that the sporadic as well as the new variant form of CJD may be tied to eating contaminated meat. If this is true, then the official data from the National CJD Surveillance Unit should show a rise of "sporadic" CJD deaths in the United Kingdom as well. And they do. The rise from fewer than ten cases in 1970 to more than fifty cases in

2002 is a definite upward trend for a disease that supposedly occurs "spontaneously" or randomly. Trying to brush aside the noteworthy increase, the CJD surveillance unit explained: "This may be attributed to improved case ascertainment."

Another set of research findings, potentially even more significant than those presented by Health Minister Dorrell in front of the House of Commons on March 20, 1996, were announced by England's Mental Health Foundation on the same day. In an exhaustive project, a group of pathologists with the foundation had examined the brains of more than 1,000 people who had died with the diagnosis of dementia and found that 19 of those brains had CJD, according to Rampton and Stauber. Only half of them had been correctly diagnosed. The remaining half had been diagnosed either as Alzheimer's disease or as some other form of dementia. "It's clear a significant number of cases are slipping through the diagnostic net," said June McKerrow, the foundation's director. "We're concerned that public health information is currently based on data which may be misleading or inaccurate."

Not only are the official CJD numbers in the United Kingdom almost certainly highly underrepresented, but a significant percentage are likely to be hidden in the numbers for Alzheimer's disease, which has now reached epidemic proportions in the United Kingdom.

The situation would not be far different across the Atlantic.

14

Prime Cuts

Watching all this from across the pond, Americans were rightfully wondering: Is there a mad cow risk in the United States? In 1989 the United States Department of Agriculture and the Food and Drug Administration put together a top scientific committee to examine the risk. Richard Marsh, the Wisconsin veterinarian, was a member of that committee. But within a year, according to the authors of *Mad Cow U.S.A.*, "this committee was completely dominated by the meat industry, the rendering industry, the sheep industry and the dairy industry." The group essentially met in secret. No representative from the public has ever been on the committee. When Marsh warned the committee of the need to stop the practice of feeding animals back to the same species, Rampton and Stauber noted, "He was essentially ignored, ridiculed, even threatened at times."

In 1990 Marsh set down his evidence in black and white in a scientific letter to the *Journal of the American Veterinary Medical Association*. At one time he believed that little, if any, rendered animal products were used for protein supplements in cattle feed in the United States. But, wrote Marsh, "I have since learned that this is incorrect, because of the recent trend of using less assimilated by-pass proteins in cattle feed. A large amount of meat-and-bone meal is being fed to American cattle, and this change in feeding practice has greatly increased the risk of bovine spongiform encephalopathy (BSE) developing in the United States." He carefully laid out the data on the mink outbreak in Stetsonville, caused apparently by rendered cattle feed, but his letter, like his symposium speeches, fell on deaf ears.

Many in the scientific community appeared cowed by the virulent attacks that had been unleashed on Marsh and they seemed unprepared to risk their own scientific careers. As if to reinforce his point, Marsh penned a second letter stating the situation was even more serious than he had first supposed. But again the National Cattlemen's Beef Association, the USDA, and other agencies ignored Marsh's concerns. The National Renderers Association (NRA) described as "ludicrous" Marsh's contention that downer cows might contain a subset infected with the scrapielike agent and might have caused the TME outbreak in Stetsonville. And the NRA dismissed Marsh's data as "anecdotal." According to reporter Andrew Nikiforuk of Toronto's *Globe and Mail*, "Prof. Marsh was vilified and denigrated by the U.S. cattle industry for his work. His grant proposals to test more cattle were routinely turned down by government."

During his prolific three-decade career, Richard Marsh had plowed a lonely furrow in the arena of veterinary research. He was a very reluctant warrior in his bid to warn the American public that a dangerous pathogen lurked in their food supply. Although he could not produce a smoking gun that the scrapielike agent was actually present in U.S. cattle, his circumstantial evidence was carefully gleaned and overwhelming in its implications. His scientific data were routinely belittled and ignored in favor of maintaining the status quo, a position driven simply by economics.

Howard Lyman, a retired rancher and an activist, described what happened to Richard Marsh at a symposium on BSE that Lyman attended at the University of Wisconsin–Madison in 1993. "It was like they walked him [Marsh] up to the gallows, put the rope around his neck and sprung the trap," said Lyman. "I believe the entire symposium was orchestrated simply to bring Dick Marsh to heel. I think it broke his heart. I think Marsh is a big teddy bear, a brilliant researcher, a wonderful human being, but he has no shell against that kind of attack. Some people when you pick on them they get tougher, others just wilt."

The "they" that Lyman was referring to is the rendering industry. Prior to the 1980s the rendering industry in the United States kept an extremely low profile. Most people did not know of its existence, and if they did, it was only in the vaguest of terms. The rendering industry was, and still is, a billion-dollar enterprise tasked with the removal of cattle that are not eaten by humans. The industry disposes of all cattle remains after death, whether sick or healthy. Over the decades the industry has found a remarkably efficient method

of recycling the parts of dead cows back into the North
American animal food chain, as well as into countless products used daily by Americans who remain blissfully unaware
of their origin.

A good summary of the system that is the rendering
industry appeared in an oft-quoted article written by reporter
Van Smith of the *Baltimore City Paper* who was asked to do a
story on a local rendering plant. "Consider these items,"
wrote Van Smith in 1995,

> the Baltimore City Police Department quarter horse
> who died last summer in the line of duty. The grill
> grease and used frying oil from Camden Yards, the
> city's summer ethnic festivals, and nearly all Balti-
> more-area . . . restaurants and hotels. A baby circus
> elephant who died while in Baltimore this summer.
> Millions of tons of waste meat and inedible animal
> parts from the region's supermarkets and slaughter-
> houses. Carcasses from the Baltimore zoo. The thou-
> sands of dead dogs, cats, raccoons, possums, deer,
> foxes, snakes, and the rest that local animal shelters
> and road-kill patrols must dispose of each month.
>
> These are the raw materials . . . which are
> processed into marketable products for high profit at
> the region's only rendering plant. . . . In a gruesomely
> ironic twist, most inedible dead animal parts, includ-
> ing dead pets, end up in feed used to fatten up future
> generations of their kind. Others are transmogrified
> into paint, car wax, rubber, industrial lubricants. . . .
>
> During a midsummer day's visit to the plant, I gag

upon first contact with the hot putrescent air. My throat immediately becomes coated with the suety taste of decayed, frying flesh.

"You picked a bad day to visit a rendering plant," [plant manager Neil] Gagnon says. . . . "By the time we get [the dead animals] they're soup. . . . Summertime is bad around here."

A load of guts, heads, and legs, recently retrieved from a local slaughterhouse, sits stewing in one of the raw material bins at the plant's receiving bay. . . . it will be fed into the "hogger," a shredder that grinds up the tissues and filters out trash, before it is deep fried in cookers charged with spent restaurant grease and blood. . . .

Blood and body fluids leak out from under the trailer gate. . . . Suddenly a hot gust of wind blows droplets of it on our bare legs. As the bloated stomachs and broken body parts slide en masse from the trailer bed to the bin, Bud [the cooker operator] shouts out, "Watch out for the splatter!" . . .

. . . [T]he plant's owner catches wind that the press has entered the property. . . . [H]e ushers us off to the adjacent sidewalk. . . . "[T]here just is no good publicity for us right now," he explains.

Smith's article created a firestorm from a public horrified to learn that not only were cows rendered along with roadkill, but that maybe even their pets were being rendered and fed back to their own pets as pet food. Such graphic articles on the rendering industry were extremely rare prior to the 1990s.

A year after Smith's exposé appeared the issue was raised again, this time before a national television audience. Richard Marsh was dying of cancer and probably was not around to see *The Oprah Winfrey Show* that was broadcast April 15, 1996. But Winfrey would see her own version of the dismissive and insulting treatment received by Marsh from the cattle and rendering industries after her show aired.

On the show Oprah looked aghast as her guest, Howard Lyman, the retired rancher turned vegetarian, declared: "A hundred thousand cows per year in the United States are fine at night, dead in the morning. The majority of those cows are rounded up, ground up, fed back to other cows. If only one of them has mad cow disease, it has the potential to affect thousands."

Oprah blinked. "But cows are herbivores. They shouldn't be eating other cows," she said with concern in her voice.

Lyman replied: "That's exactly right, and what we should be doing is exactly what nature says. We should have them eating grass, not other cows. We've not only turned them into carnivores, we've turned them into cannibals."

There was an audible gasp from the audience. This was the first time in history that this subject had been discussed before a large, nationwide television audience.

The American public had never guessed the dirty little secret of the rendering industry and the cattle industry that for economic reasons had completely distorted the food chain of cows and had turned these placid, docile creatures into cannibals. Reaction was swift and immediate.

The cattle industry watched as cattle futures dropped like a stone following the airing of the Oprah show. The beef industry pulled about $600,000 in network advertising. So great was the furor from the cattle industry that Oprah was forced to schedule a second show the following week. As activists Rampton and Stauber wrote: "The follow-up show, which aired a week later, featured a ten-minute one-on-one exchange between a cowed Oprah Winfrey and Gary Weber (the National Cattlemen's Beef Association representative). . . . As Weber issued reassurances, Oprah uttered weak half apologies that seemed as though they were being forced through gritted teeth." But the industry was not appeased. They made Winfrey's life miserable with a barrage of expensive legal maneuvers.

After six years of litigation involving two separate lawsuits, Oprah was eventually vindicated in September 2002. U.S. District Judge Mary Lou Robinson finally rendered all claims against the TV diva null and void, but the frontal assault by the cattle lobby on her had taken its toll. The cattle lobby had made its point. If you want to criticize beef, you will do so at very high legal and personal cost.

15

The Alzheimer's Nightmare

The Centers for Disease Control insist there is no need for worry; the incidence of CJD in the United States is extremely low. Only one in a million people get Creutzfeldt-Jakob disease in this country. In the twenty-year period between 1979 and 1998, the official figures show that there have only been 4,751 CJD deaths in the United States. That's fewer than 250 deaths per year. If mad cow disease were a cause to worry, they argue, the number of CJD cases would be much higher. Besides, as the United States Department of Agriculture likes to emphasize, the true measure of the transmission of prions from cattle to humans is a supposedly different disease known as variant CJD. So-called sporadic CJD, which accounts for all 4,751 cases in the United States, has nothing whatsoever to do with eating tainted beef, it says.

But scientific research over the past three years shows that these reassurances from the United States Department of Agriculture and the Centers for Disease Control are deeply flawed. The evidence now suggests that prion diseases in humans are much more widespread than previously thought. This evidence comes from several directions, but it all points to the same chilling conclusion—that the situation is very likely much worse than officials have been willing to acknowledge.

To begin with, the Centers for Disease Control and the United States Department of Agriculture statistics on CJD are far from authoritative. The simple fact is that there is no mandatory reporting of CJD cases in many states, so many CJD deaths may be slipping through the cracks, and no one knows otherwise. When a confident-sounding spokesperson trots out the reassuring "fact" that only one in a million people get CJD, what they are actually telling us is that they have no idea how many CJD cases there are *because no one has carefully searched for them*. To think that in 2004 there is no way of accurately determining the number of CJD cases in the United States belies belief.

In researching the epidemiology and statistics of CJD in this country, I was surprised to learn that much of the data gathering is conducted by volunteer organizations, mostly composed of individuals who have had a loved one die of CJD. Is this the way modern medicine should operate? Neither the Centers for Disease Control nor the National Institutes of Health (whose annual budget now approaches $30 billion) has stepped up to the plate. The National Prion Disease Pathology Surveillance Center has a standing request for the brains of people who die of CJD, but their budget is pitiful and most people do not know of the organization's existence.

Underlying the reporting problem is the enormous difficulty in correctly diagnosing CJD. The fact of the matter is that when faced with the profound overlap of symptoms— dementia, muscle jerking/spasm, and initial memory problems—between CJD and Alzheimer's (or other dementia), many physicians have difficulty distinguishing between the two diseases. Their initial diagnosis of Alzheimer's, rather than CJD, in many cases occurs simply because physicians are more familiar with Alzheimer's.

Currently, the only way to definitively diagnose CJD requires drilling into the dead victim's skull on an autopsy slab, extracting parts of the victim's brain, and looking for the characteristic holes in the brain cells with a microscope. A big roadblock to adopting widespread autopsies of suspected CJD victims is the profound reluctance on the part of medical professionals to conduct autopsies on suspected cases of CJD, as there have been several documented cases published in the medical literature in which CJD was transferred from surgical instruments to other people. "There is clear evidence from case reports in humans and animal models that prion diseases can be transmitted via stainless steel instruments," stated the *British Medical Journal* in 2001.

"Prion disease researchers say their efforts to track and discover new strains of the disease are hampered by a dramatic decline in the number of autopsies in the United States," wrote Elizabeth Williamson in a *Washington Post* article published in January 2004. "Nationwide, autopsies are conducted in five percent of non-crime-related deaths, compared with 35 percent in the 1960s, according to the prion disease committee at the Institute of Medicine of the National Academies, which advises the federal government on public health issues.

"The congressionally chartered institute said that examining brain tissue is the only sure way to pinpoint a prion disease. Yet 'at least half of the estimated total number of deaths caused by [prion diseases] in the U.S. are not autopsied and confirmed by laboratory examination.'"

A further complication is that most health insurance companies in the United States will not pay for autopsies, providing little or no incentive for coroners to conduct them. This means that a grieving family that has just lost a loved one through the horrors of CJD would have to pay out upwards of $1,500 for an autopsy to confirm death by CJD. How many families are going to do that? The result is that we have no idea what the real CJD numbers are. "If we don't do autopsies and we don't look at people's brains," Dr. Laura Manuelidis of Yale University told a reporter for United Press International, "we have no idea about what is the general prevalence of these kinds of infections and (whether) it is changing."

Laura Manuelidis is one of the top neurodegenerative disease scientists in the world. In 1989 she and her husband, Elias, who died three years later, published the results of a study that would make a mockery of the Centers for Disease Control statistics for CJD prevalence in the United States. At Yale University, the husband-and-wife team had examined the postmortem brains of forty-six Alzheimer's disease patients and found six cases, or 13 percent of the cases, where the patients had died of CJD. The six patients had been *misdiagnosed as Alzheimer's patients.*

"Neurodegenerative disease and Alzheimer's disease have become a wastebasket" for mental illness in the elderly that is difficult to diagnose conclusively, said Laura Manuelidis in 2003. "In other words, what people call Alzheimer's now is more broad than what people used to call it, and that has the possibility of encompassing more diseases—including CJD." Dr. Manuelidis has been able to cite unpublished evidence going back almost twenty years showing that misdiagnosis of Alzheimer's may be masking an epidemic of CJD in North America.

The essential findings of the Manuelidis study were confirmed in a second study, this one conducted at the University of Pittsburgh and published in the journal *Neurology,* also in 1989. The Pittsburgh study examined the brains of fifty-four presumed Alzheimer's patients and found three cases of CJD, or 5.5 percent of the cases. The Pittsburgh and Yale studies suggest that somewhere between 5.5 percent and 13 percent of Alzheimer's diagnoses are really CJD. Why is this worrisome? Because the current epidemic of Alzheimer's disease in the United States in 2004 threatens to engulf the entire United States healthcare system.

We are told that Alzheimer's disease is an old people's disease. Indeed, nowadays "dementia" and Alzheimer's disease are considered a normal consequence of growing old. But are they? Consider this. In 1979, according to official statistics of the Centers for Disease Control, 653 people died of Alzheimer's disease in the United States. In 1991, that figure leapt to 13,768. And by 2002 more than 58,785 people died of Alzheimer's disease in the United States. In other words, in a period of twenty-four years there has been an 8,902 percent increase in deaths from Alzheimer's disease in the United States. Clearly we have a huge epidemic on our hands.

But this trend is nothing to worry about, according to a 1996 report issued by the Centers for Disease Control: "The increasing trend may reflect improvements in diagnosis, awareness of the condition within the medical community, and other unidentified factors rather than substantial changes in the risk of dying from Alzheimer's Disease." Currently, about 4.5 million people in the United States have Alzheimer's disease and will die from it, and projections show that this figure will more than double by 2020. It boggles the mind that almost a 9,000 percent increase in two dozen years in any disease can be airily dismissed by changes in disease classification and other vague nostrums.

Even worse, it's likely that the annual rates of Alzheimer's disease itself, as published by the CDC, are being underreported. In conversations with several Alzheimer's groups in the United States, I learned that since reporting of Alzheimer's is not mandatory, there is a large uncertainty factor in determining the actual number of cases. So the true figures for Alzheimer's disease, as with CJD, are essentially unknown.

Not surprisingly, then, there is also considerable uncertainty nationally regarding the precise numbers of "early onset" Alzheimer's disease cases in the United States. Knowing what the real numbers are for early-stage Alzheimer's is important because if more and more people are coming down with the early stage of this disease, then it becomes even more urgent to conduct autopsies on these people to make

Dr. Carleton Gajdusek with Fore tribe children in New Guinea. Gajdusek investigated kuru in the Fore tribe and won the Nobel Prize in 1976. (PEABODY ESSEX MUSEUM, SALEM, MA)

Dr. Joseph E. Smadel, C
leton Gajdusek's boss at
National Institutes of Hea
was one of the most powe
figures in Washington med
circles in the 1950s and ea
1960s. (WALTER REED ARCHIVES)

Dr. Richard Marsh discovered evidence for BSE in the United States in the 1980s. (UNIVERSITY OF WISCONSIN AT MADISON)

This dramatic wool loss developed in a sheep with scrapie in a single week. (Dr. Michelle Crecheck, APHIS)

A typical case of a deer with subtle symptoms of chronic wasting disease, according to Dr. Elizabeth Williams. (Dr. Elizabeth Williams, University of Wyoming).

Millions of suspected BSE-infec
cattle were killed and their carcas
burned in the United Kingdom d
ing the 1990s. (PA PHOTOS)

Agriculture Minister John Gumr
and his daughter Cordelia eat ha
burgers to assure the British pub
that beef is safe. Shortly afterw
scores of people began dying of C
(PA PHOTOS)

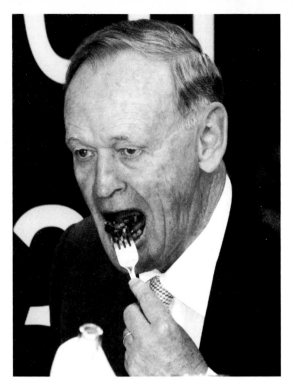

Canadian Prime Minister Jean Chretien eats beef in a restaurant in Ottawa in May of 2003 to assure the Canadian public that eating beef is safe. (DAVID CHEN, CANADIAN PRESS)

A drooling elk with chronic wasting disease. (DR. ELIZABETH WILLIAMS, UNIVERSITY OF WYOMING).

above: A cow mutilated in Montana in 1975; its brain, lips, and tongue were removed. Was this part of a covert disease-monitoring operation? (Captain Keith Wolverton)

below: A veterinarian determined that sharp instruments had been used to mutilate this cow and many like it in Montana. Between 1974 and 1977, more than sixty animals were found mutilated around Great Falls, Montana. (Captain Keith Wolverton)

Map of northwestern Saskatchewan, Canada, showing the locations of cattle mutilations (circles), CWD (stars), BSE (square), and scrapie (triangle). Is the overlap coincidental?

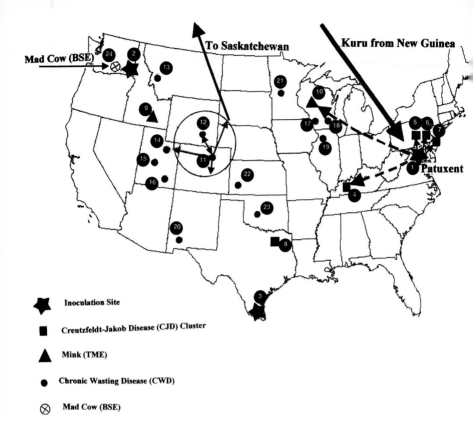

Mad Cow (BSE)

To Saskatchewan

Kuru from New Guinea

Patuxent

★ Inoculation Site

■ Creutzfeldt-Jakob Disease (CJD) Cluster

▲ Mink (TME)

● Chronic Wasting Disease (CWD)

⊗ Mad Cow (BSE)

The big picture. Kuru brains from New Guinea were injected into multiple species at Patuxent Wildlife Reserve (1). If the infectious disease escaped it may have been responsible for nearby CJD clusters in Kentucky via infected squirrels from (4), LeHigh Valley, PA (5), Allentown, PA (6), Cherry Hill, NJ (7). Cattle were inoculated with prions at Pullman, WA (2), and Mission, TX (3). An outbreak of TME occurred in Blackfoot, ID (9), with unknown origin. But was the inoculation at Patuxent also responsible for the TME outbreak in Stetsonville, WI (10), via contaminated cattle? And did a transfer of wildlife or sheep to Colorado in the early 1970s contribute to the spread of CWD? We will never know. CWD erupted in Fort Collins, CO (11), and in Sybile, WY (12), in the late 1960s, and the epidemic spread into Wyoming, Nebraska, and through Colorado (large circle). Subsequent outbreaks occurred in captive animals in Phillipsburg, MT (13), and in wild deer at three locations in Utah (14-16), and in Wisconsin (17,18) and Illinois (19). CWD was exported to Saskatchewan from South Dakota. An unexplained jump of CWD into White Sands, NM (20), occurred. CWD has been found in captive cervids in Kansas (22) and Oklahoma (23). In December 2003, the first official case of BSE was announced in Mabton, WA (24).

sure they are not actually dying of CJD instead. According to my conversations with Alzheimer's organizations, the conservative estimate for early-onset Alzheimer's varies between 5 and 10 percent of total Alzheimer's cases. The early-onset subgroup is defined as those patients who succumbed to symptoms before age sixty-five. Therefore, if the figure totalling 4 to 5 million Alzheimer's disease patients is taken as accurate, between 200,000 and 400,000 of those patients showed symptoms before age sixty-five. Whether the number of "early onset" patients is, or has been, on the rise cannot be ascertained by studying currently available data.

The situation is similar in the United Kingdom. Given that the U.K. has seen about 150,000 cows succumb to BSE since that fateful day in April 1985 when Jonquil first starting showing symptoms, it could be predicted that rates of Alzheimer's disease would have risen there, too. And that is precisely what the statistics show. In 1979 less than one person in every 100,000 died of Alzheimer's disease. By 1996, eighteen years later, that figure had jumped *1,900* percent to twenty cases per 100,000. And the figure is still rising. "There are over half a million people with Alzheimer's disease in the UK today," according to the Alzheimer's Research Trust in Great Britain. Other estimates put the current U.K. figure closer to one million. The Alzheimer's Disease Society of the United Kingdom estimates that out of the 700,000 patients with the disease in Britain, about 17,000 people under the age of sixty-five suffer from the illness. In the United Kingdom, dementia and Alzheimer's disease are overlapping populations, so there is some confusion regarding the actual numbers. But there is no doubt regarding the trend of

Alzheimer's disease in Britain. A catastrophic increase in Alzheimer's shows up in both the United States and the United Kingdom.

As global research into the genetic susceptibility to CJD and Alzheimer's has intensified, one recent finding stands out. It turns out that the people with the same prion mutation ("Met/Met 129" or "Val/Val 129") have an increased risk for dying of *either* CJD or early-onset Alzheimer's. Furthermore, a remarkable paper published in the journal *Neurology* in July 2004 specifically suggests the involvement of prions in early-onset Alzheimer's disease. This recent research provides yet more overlaps between CJD and Alzheimer's disease.

Finally, it is worth noting the work of Dr. Hugh Hendrie, a professor of psychiatry at the University of Indiana, and his colleagues who compared rates of dementia and Alzheimer's disease between elderly people in Ibadan, Nigeria, with an equivalent group living in Indianapolis, Indiana. The overall prevalence of dementia and Alzheimer's disease in Nigeria were 2.29 percent and 1.41 percent, respectively, which was "much lower" than the 8.24 percent and 6.24 percent found for African Americans in Indiana. "The differences in the frequencies of vascular risk factors between Nigerian subjects and African Americans," noted Hendrie, "would suggest involvement of environmental factors in the disease process."

Suffice it to say, it seems difficult to imagine that the extraordinary rise of Alzheimer's disease since 1979 can simply be explained away by improvements in diagnosis of the disease and by other "unidentified factors." I suggest that one of the unidentified factors in the Alzheimer's equation may be an unidentified infectious agent within the human food chain. If the 4 to 5 million cases of Alzheimer's disease in the United States is indeed masking a hidden epidemic of misdiagnosed CJD, the Yale and Pittsburgh studies suggest that the true CJD numbers may be more on the order of tens or hundreds of thousands, far more than the 4,751 cases the Centers for Disease Control has counted. These numbers in the tens or hundreds of thousands describe an alarming CJD epidemic that currently lurks beneath the radar of the American medical establishment. This begs the question, if CJD is actually much more common than believed, where is it coming from?

16

Clusters

A central plank in the assurances coming from the Centers for Disease Control and the United States Department of Agriculture is that sporadic CJD has nothing whatever to do with eating tainted meat. Only variant CJD is tied to eating meat, so their argument goes, and there has never been a case of variant CJD so far in the United States. Ergo, they say, eating beef is safe in the United States. But again the data suggest otherwise.

The differences between sporadic CJD and variant CJD are not as clear-cut as medical authorities would have us believe. When researchers at the University of London reported in 2002 that they had injected BSE-infected brain tissue into several groups of transgenic mice, or mice genetically engineered with human prion protein, they found one group of mice displayed, as expected, the brain lesions of variant CJD. But the shocker came when they looked at a second group, which they had also injected with BSE. The second group had brain lesions that exactly mirrored *sporadic* CJD. This was not supposed to happen. The direct implication from the experiment is that BSE can also cause symptoms of sporadic CJD. If BSE can cause both variant CJD and sporadic CJD in humans, the whole balloon of assurances from United States Department of Agriculture and the Centers for Disease Control simply collapses.

If the more common form of CJD, sporadic CJD, is also linked to BSE, as the University of London study suggests, then animal-to-human transmission is easier than previously assumed. In an interview with the BBC, Professor John Collinge, the senior author of the study, said: "When you counsel those who have the classical sporadic disease, you tell them it arises spontaneously out of the blue. I guess we can no longer say that." A month later, Yale's Laura Manuelidis told UPI reporter Steve Mitchell: "Now people are beginning to realize that because something looks like sporadic CJD, they can't necessarily conclude that it's not linked to [mad cow disease]." Indeed, a team led by Adriano Aguzzi of the University Hospital of Zurich, Switzerland, recently reported in the *New England Journal of Medicine* that there was a sudden twofold increase in sporadic CJD figures in 2001 in Switzerland and suggested BSE might be to blame. Switzerland has a relatively high rate of confirmed BSE cases compared with the rates of other countries in Europe. (It has been suggested the reason why Switzerland has such a high incidence of BSE is because it is the only country honestly reporting BSE cases.)

Perhaps the most important and troubling finding that bears on the so-called spontaneity of sporadic CJD cases is that in spite of the inefficient system of reporting, persistent descriptions of CJD clusters exist in the United States.

Janet Skarbek, a dynamic, very articulate, thirty-six-year-old professional woman, lives in Cinnaminson, New Jersey, a suburban community about thirty miles southwest of Trenton. Janet's friend Carrie Mahan was only twenty-nine when, in January 2000, she became ill. It started off with fatigue and then rapidly progressed to hallucinations, and when she was admitted at the University of Pennsylvania Medical Center in Philadelphia, her condition deteriorated rapidly. On February 24, 2000, Carrie died, and a subsequent autopsy showed her brain was so shot full of holes that it looked like Swiss cheese. Her doctor suspected CJD even before the autopsy, although twenty-nine seemed awfully young to die from CJD. Victims of "sporadic" CJD are usually much older.

Pat Hammond is Janet's mother and she had hired Carrie to work at the Garden State Race Track in Cherry Hill, New Jersey. Janet's home is no more than ten miles north of the racetrack. In June 2003, while randomly looking through the obituaries, Janet noticed that a Carol Olive had also died of CJD. What floored Janet was the second section of the obituary. Carol had worked at the Garden State Race Track. After finding Carol's linkage to the racetrack, "I nearly fell over," Janet told *The New York Times*.

Later that day, Janet confirmed it with her mother, who remembered Carol. She was astounded that two people who worked at the racetrack had died of the same disease, even though "sporadic" CJD was a one-in-a-million kind of disease. After a very restless night Janet searched the obituary databases in the area. Lo and behold, she found that John Weber, who had lived nearby in Pennsauken (which lies directly between Cinnaminson and the racetrack) had also died of CJD in 2000. Like the gutsy person she is, Janet decided to take a chance and call the mourning relatives of John Weber. William Weber answered. He said his brother had a season pass to the racetrack. "Skarbek dropped the phone. Weber was her 'eureka' moment," reported *The New York Times*.

Through perseverance, Skarbek went on to collect over a dozen cases of CJD, all associated with eating in the cafeteria at the Garden State Race Track. The CDC response was luke-warm at best. They seemed more interested in making sure the word *cluster* was not used in connection with the investigation. According to a follow-up report in Britain's *Independent*: "Ms. Skarbek has given the CDC, as well as the New Jersey Department of Health, documentation that allegedly demonstrates that as many as thirteen people who have died in her corner of New Jersey and in nearby Pennsylvania and who were diagnosed with sporadic CJD, were all either workers at the closed race track or were patrons there and at its cafeteria. Another six deaths in the region seem to fit the same pattern, she says, although she is still trying to confirm whether at least two of those victims had connections with the track."

When I spoke with Skarbek in mid-April 2004, she told me she had confirmed fifteen CJD cases of people who had eaten at the track, and preliminary indications are that the majority of those fifteen deaths seemed to track to a single restaurant. (There were three restaurants and several concessions at the racetrack.) In addition to those fifteen confirmed cases, there were a few more who had either died but had not yet been autopsied, or who were still dying and whose brain biopsies looked suspicious. She told me because of the publicity from *The New York Times* article, she gets calls every day from people with families who have died of CJD or even from families who still have a loved one in the terminal stages of CJD.

In spite of what initially looked like stonewalling from the health authorities, a big break occurred in early April 2004 when Janet got the backing of the two powerful New Jersey senators, Frank Lautenburg and John Corzine, who expressed their concern to the Centers for Disease Control and urged them to conduct a "thorough investigation," according to the *New Jersey Courier Post*. South Jersey Congressman Rob Andrews had made a similar plea two months earlier after meeting with Skarbek in January. At the end of March Andrews had received a letter from Dr. Julie L. Gerberding, director of the CDC, in which she called Skarbek's allegations "extremely unlikely."

A central plank in the dismissal of the evidence by the CDC is that the restaurant in question at the Garden State Race Track had almost a dozen separate meat suppliers. So, the CDC argues, the alleged tainted meat would have to have been present at the restaurant and served to a lot of customers for a very short time. Hence, according to the CDC, the four-year period that Skarbek alleges that the victims might have visited the racetrack seems highly unlikely. Given all the brouhaha about the impossibility of the alleged tainted meat coming in from so many different sources, it seems amazing that contamination of the premises itself does not appear to have been considered or investigated by the CDC. Or by the New Jersey Department of Health.

Since the lethal transfer of prions from surgical instruments to other people has been shown to occur on several occasions, authorities should certainly have considered whether prions from contaminated meat could have infected the food preparation area and the knives used to cut the meat. The hypothesis that large-scale contagion of the food preparation area over a few months' period between 1988 and 1992, when all victims seem to have eaten at the restaurant, seems an obvious forensic path to follow. Since it is well documented that the infectious particles in scrapie remain active and infectious in the soil for years, it does not seem out of the question that prions from contaminated meat might have contaminated the food preparation area or its utensils over a period of months. There is a possibility that the prions involved survived standard dishwashing procedures in the restaurant and could be continuously infectious over a period of time.

The CDC's apparent failure to adequately investigate this possibility seems foolhardy. When I mentioned that possibility to Skarbek, she said that she had been told that some of the kitchen equipment from the possible "hot-zone" restaurant in question had since been sold to several restaurants in the New Jersey area. If any of that equipment had actually been contaminated at the restaurant, it might now be spreading that contamination among several restaurants in New Jersey.

The New Jersey health authorities launched an investigation into the purported cluster and in May 2004 published a report, which was echoed at the federal level, saying that because the CJD found at the racetrack was sporadic, not variant, CJD, there was nothing to worry about since sporadic CJD was not tied to eating BSE-tainted meat. This statement directly contradicts the already published scientific evidence from Professor Collinge's group that BSE transmission to transgenic mice expressing the human prion protein may take the form of *either* sporadic or variant CJD. The New Jersey health authorities and the CDC were telling the public exactly what Collinge said you can't say with certainty about CJD. Skarbek has told me that Collinge has taken a strong interest in her case for the New Jersey cluster. The Japanese government, which continues to enforce a ban on the import of United States beef as of May 2004, and the media are also following this case with intense interest, according to Skarbek.

And New Jersey is not alone. Clusters of CJD have also been reported in Pennsylvania in 1993, Florida in 1994, Oregon in 1996, Texas in 1996, and New York between 1999 and 2000. The apparent clustering of CJD raises the question: If CJD arises "spontaneously" with no known cause, as we are told, why should it cluster? Clustering implies a nonrandom cause for the disease.

One disturbing possibility suggested by this clustering is that CJD is infectious. Indeed, new concerns and questions about the potential infectiousness of CJD were raised in February 2003 with the report from Italian scientists showing that large concentrations of prions were found in the nasal cavities of CJD victims. In their paper, published in the *New*

England Journal of Medicine, the authors suggest that their finding holds important implications not only for diagnosing CJD, but also for the risk of infection from living patients. One of the more worrying implications from the Italian study is that maybe prions are not just in the brains of infected people. If they are in the mouth or nasal cavity, does that mean that CJD is more infectious than thought? And this is a disease that is 100 percent fatal. There is no cure.

Support for this chilling scenario came in November 2003, when a team led by Dr. Adriano Aguzzi, one of the world's experts on CJD, found pathologic prions in the spleen and muscle of approximately one third of the CJD victims they examined. Their paper, which also appeared in the prestigious *New England Journal of Medicine,* implies that prions in humans are not confined to the brain and may be much more widely transmissible as a result of surgery. The Aguzzi study also holds major implications for prion diseases in deer and cattle. If prions were found in the muscle of one-third of humans infected with CJD, there is a strong implication that prions can be found in the muscle of cattle that have mad cow disease, even cattle that do not have symptoms of BSE. A second study, this one from Germany and published in May 2004, shows substantial accumulation of prion proteins in the muscle of hamsters that were orally fed prions, and lends ammunition to the claims that prions in muscle are a significant concern. These findings completely undermine the bland assurances that are currently emanating from both the Centers for Disease Control and the United States Department of Agriculture. These facts are known in the scientific literature, but still are not being addressed by the beef lobby.

In February 2004, Dr. Aguzzi dropped another bomb-shell. He showed that the variant form of CJD could be trans-mitted via blood transfusion. This news sent alarms through the medical establishment on both sides of the Atlantic because it provided yet another nightmarish scenario for the possible spread of CJD. Aguzzi's finding was followed a month later by a British government announcement banning blood donations. "Thousands of people have been banned from donating blood because of fears over the human form of BSE," the BBC reported in March 2004. "The ban, announced in the Commons by Health Secretary John Reid, applies to all those who have had blood transfusions since 1980."

In another sobering report, released in the *Journal of Pathology* in May 2004, researcher David Hilton and col-leagues from the Derniford Hospital in Plymouth, England, documented the detection of variant CJD prions in a random sample of surgically removed tonsils and appendixes, 60 per-cent of which were in patients in the 20 to 29 age range when the operations occurred. From this finding, the researchers concluded that the incidence of variant CJD in the "healthy" general population numbered in the several thousands. On the same day, the headline in *The Independent* stated: "Scien-tists Fear Hidden Epidemic of vCJD." This piece of bad news belied the widespread assurances from the Health Ministry that mad cow–related CJD was on the wane in the UK.

As this finding suggests, humans and animals may in fact have CJD and BSE, respectively, without showing any symp-toms of the disease at all. Professor John Collinge and his coauthor, Andrew Hill, dealt with this "subclinical" state of prion disease, where the patient shows no overt signs of the disease, in an article published in *Trends in Microbiology* in December 2003. The authors cite abundant evidence that cattle, humans, and other animals can harbor relatively high levels of prion infection without necessarily displaying symptoms and they warn of the possibility of "carriers" that can infect other people or animals. This, the authors state ". . . should also encourage consideration of the possibility that other species (such as sheep, pigs and poultry) exposed to BSE prions by contaminated feed, might be able to develop" the disease while showing no signs of it. Collinge and Hill then conclude: ". . . it must be considered possible, if not probable, that BSE passaged in animals other than cattle will retain pathogenicity for humans."

Strip away the rather dry, technical language of this extraordinary statement, and what Collinge and Hill are really saying is that we cannot trust poultry or pork if they have been fed contaminated feed, and that poultry and pork may be capable of passing on BSE to humans. This statement foreshadows a much deeper and more calamitous possibility that not only is the human food chain compromised from beef, it is also compromised from chicken and pork. If animals can harbor high levels of infectious prions in their brains without showing any clinical signs of disease, state Collinge and Hill, then "prion infection as opposed to disease should be tested in apparently healthy cattle following slaughter to investigate whether significant levels of 'silent BSE' (sub- or preclinical) are present."

What this all boils down to is that it is almost useless to just test dead animal (or human) brains for the disease. You have to look for the prions themselves and only then will you get a true picture of the extent of infection. Only by testing for prion infection is it possible to uncover the extent of the silent catastrophic iceberg of a massive prion epidemic throughout the food chain that lurks beneath the cant, the propaganda, and the bland assurances from the beef industry that eating beef is still safe.

And cattle may not be our only source of worry. There is yet another major player in this epidemic, one whose critical role in spreading the disease is still being explored.

17

Mad Deer

Elizabeth Williams was a young graduate student at the wildlife research facility at Colorado State University in Fort Collins, a world-class facility of research into nutrition and diseases of "non-domesticated ruminants" including wild deer, elk, and bighorn sheep. While Williams was working with Dr. Stuart Young, an expert in animal infectious disease, the facility housed a number of captive mule deer.

In 1967 Williams and her coworkers noticed that one of the mule deer was behaving oddly. It was listless and seemed depressed. She kept an eye on it and she noticed that over time it began to drool and to urinate excessively. The animal had also stopped eating, though it would grind its teeth almost continuously, and within several weeks it became physically emaciated. A few weeks later the animal was dead.

Other captive mule deer at the facility would suffer the

same fate. Between 1974 and 1979, the disease, which always followed the same course, affected fifty-three of sixty-seven mule deer and one black-tailed deer. It did not seem to matter what time of the year the mule deer fell ill; the symptoms appeared regardless of the season. Williams and her coworkers thought the animals were perhaps poisoned or maybe suffering from dietary insufficiency. Occasionally, feed contamination or deficiencies lead to really bizarre symptoms.

As Williams and her colleagues studied the sick animals, it became obvious that the animals experienced periods of lack of awareness during which they were vacant and would be seen standing alone in the deer run. The disease was inexorable and 100 percent fatal. It did not discriminate among its victims. It affected animals whether raised from infancy by hand or captured in the wild. It brought down males, females, and castrated deer, but never affected animals younger than two and a half years old. Several years would pass before pathology studies were conducted on their brains. In the interim every possible nutritional variable was tested without success. Still the animals kept dying.

At the same time as the Colorado episode, a similar outbreak was occurring at a facility in nearby Wyoming—the Sybille Wildlife Research and Conservation Education Unit in the Game and Fish Department at the University of Wyoming. The facility is situated some fifty miles northeast of Laramie, Wyoming, through the Laramie Mountains. Deer and other animals were constantly transported between the Colorado and Wyoming facilities, according to Dr. Terry Spraker, who conducted research at both facilities.

The first pathologies of sick animals from Colorado and Wyoming revealed the true nature of the disease. In the gray matter of the deer brain, Williams saw the classic "spongiform" changes, the holes in the brain. In her 1980 paper, which appeared in the *Journal of Wildlife Diseases,* Williams named this scourge chronic wasting disease (CWD). The affected deer were in the center of a cornucopia of wildlife that either passed through the deer enclosures or grazed on the same grounds. "From time to time," wrote Williams, "a few young adult deer trapped in the wild have been added to the permanent captive populations. Within the facilities, deer have had irregular and discontinuous contact with other wild ruminants [elk, white-tailed deer, pronghorn antelope, Rocky Mountain bighorn sheep, and mouflon] and with domestic cattle, goats, and sheep. In addition, other feral mammalian species [mice, rabbits, raccoons, skunk, and several species of ground squirrel and dogs and cats] either reside within or traverse the facilities pens."

Once the diagnosis of spongiform disease had been made in the mule deer, Dr. Williams and her team quickly contacted the researchers at Patuxent, who, under the supervision of Drs. Gibbs and Gajdusek, were continuing to inoculate a huge variety of animals with human-derived kuru and CJD prion material. At the moment, there is no smoking gun in terms of the transfer of inoculated wild animals from Patuxent to Colorado in the 1960s, nor is there any evidence that escaped infected animals from Patuxent could have made their way to Colorado. We do know that during the 1960s and 1970s as part of a national organization, there was a transfer of wildlife facilities from Patuxent to Denver, Colorado. Whether this transfer had anything to do with the introduction of prions into the area around Colorado State University, we can only guess. Nevertheless, it has not been ruled out.

The origin of this new disease is officially still a mystery. "The origin of CWD is not known and it may never be possible to definitively determine how or when CWD arose," said Williams during a conference presentation in 2002. "Scrapie, a TSE of domestic sheep, has been recognized in the United States since 1947, and it is possible, though never proven, that deer came into contact with scrapie agent either on shared pastures or in captivity somewhere along the front range of the Rocky Mountains, where high levels of sheep grazing occurred in the early 1900s."

Of course, if there were sheep at Colorado State University in the late 1960s, if those sheep had scrapie, and if the containment facilities there were not what they were cracked up to be, it would be a lot easier to explain the outbreak of CWD at the wildlife facility. I asked Dr. Terry Spraker, a renowned wildlife researcher who was at the Colorado facility at the time, and Dr. Williams whether any sheep were kept captive at the facility. They told me that sheep were indeed kept in a "BL3" (bio-level 3) facility at Colorado State University in the late 1960s. (Bio-level 3 means the animals were quarantined and all food and water tightly controlled with air filters and personnel decontaminated when entering or leaving the facility.) These sheep had received "vibrio vaccines" and were kept in completely different research quarters. Vaccination research programs against sheep vibriosis were widespread and common in the United States in the mid-1960s. When I

asked Drs. Williams and Spraker whether any of these animals had come down with scrapie, they were unsure.

But there is evidence that the sheep did not have scrapie.
During the late 1960s, a Colorado State University (CSU)
graduate student named Gene Schoonveld began working on
his master's thesis, a nutritional study of mule deer aimed at
determining why they didn't digest the alfalfa and hay given
them during harsh winters. In addition to the test animals,
the wildlife facility also provided Schoonveld with about fifty
deer from the wild, which were held in a separate enclosure,
and, for purposes of comparative anatomy, some domestic
sheep, which were held in the enclosure with the deer.
Schoonveld, now a Division of Wildlife biologist, told the
Rocky Mountain News in 2002 that the deer "were in close
proximity of the sheep for long periods of time and it was
among those animals that the symptoms of CWD first
showed up. Soon after they were together, adult deer started
showing signs of CWD—abnormal behavior, excessive
drinking and urinating, emaciation, excessive salivation,
stumbling, trembling and depression before they died."

During Schoonveld's three-year study, an average of one
deer died of the disease each month. Necropsies by the veterinary school concluded the animals had died of "enteritis,"
which is an inflammation or infection of the intestinal tract.
But as the *Rocky Mountain News* in-depth investigation noted:
"The brain was not suspected. A number of people now recall
that the sheep in the corrals at CSU had scrapie, but no documentation has been found to prove it."

Schoonveld doesn't doubt it. "There were a number of deer
projects going on at the time and deer were coming in from

the wild that may have been infected, and we were trading deer with Sybille (the Wyoming Game and Fish Department's Sybille Research Unit, near Wheatland, Wyoming), and so it's impossible to say for sure how it got started," Schoonveld told the newspaper. "But my guess as a biologist is those sheep had scrapie, and in close confinement—something that they wouldn't do out in the wild—it jumped to deer and infected them. The deer then spread it among themselves."

Dr. Elizabeth Williams is rightly considered the matriarch of CWD research in the United States. She discovered that CWD is a prion-associated neurodegenerative disease and is a part of the larger family of fatal conditions that includes bovine spongiform encephalopathy (BSE) in cattle, scrapie in sheep, transmissible mink encephalopathy (TME) in mink, and sporadic and variant Creutzfeldt-Jakob disease (sCJD or vCJD) and kuru in humans. Her research over the years has greatly extended the available scientific information about CWD. For example, in a brilliant paper published in 2003, Williams showed that CWD was highly infectious and was spread by horizontal means in deer and elk, both wild and captive. Horizontal means the disease is spread from animal to animal; vertical transmission indicates the disease is spread from mother to offspring. Williams wrote: "Our results indicate that horizontal transmission is likely to be important in sustaining CWD epidemics."

One very plausible theory is that the facility at Colorado State University (and its sister facility in Wyoming) became the center for exporting CWD into the wilds in Colorado, and from there, nationwide. This could have happened sev-

eral ways. First, as Dr. Williams's original 1980 paper makes
clear, the CWD-infected deer were in open runs and wildlife
passed through those runs on a routine basis. Wild deer and
elk had access to the CWD-infected animals simply by
touching noses across the wire fences of the facility. Second, it
took several years of painstaking research by Williams and
colleagues to rule out dietary and other causes for the new
mysterious disease that had erupted in their mule deer. During that window of time, several of the infected animals were
given to the Denver Zoo, which in turn gave an infected deer
to a zoo in Toronto, Canada. Dr. Ian Barker, a veterinarian
from the University of Guelph, Ontario, diagnosed the mule
deer in 1978. The infected animal died at the zoo in Toronto.
Third, in the late 1960s and early 1970s, some of those
CWD-infected animals were sold out of the CSU facility to at
least one farm in South Dakota, and some were transferred to
the wildlife research facilities in Wyoming. Thus, whatever
the means by which CWD originally entered the CSU facility, a number of factors point to the wildlife research facility
at CSU as the epicenter of the nationwide epidemic of CWD.
This is not to suggest any culpability, since at the time of
many of the transfers CWD was not even recognized as a
highly infectious prion disease.

Once established at the Colorado and Wyoming facilities,
it was not long before CWD began to spread within Colorado
and Wyoming. CWD was diagnosed in captive mule deer
and black-tailed deer in Wyoming in 1979, presumably as a
result of transfers of infected animals back and forth from
Colorado, and that same year the disease was found in captive
Rocky Mountain elk. Only a couple of years later, in 1981,

the alarm bells began to ring when CWD was found in free-ranging elk in north-central and northeastern Colorado. The disease, of course, had been in the wild for several years. In the following years CWD spread inexorably outward from the CSU wildlife research facility. A map of the distribution of forty-one cases of CWD in mule deer, six cases in elk, and two cases in white-tailed deer that was published in 1997 by Dr. Spraker and colleagues shows that all CWD cases occurred within a 100 km radius of the initial outbreak at Colorado State University. According to Dr. Spraker, "affected animals were either found dead, euthanized (killed) in the field by gunshot, or captured alive and euthanized via gunshot . . . two bucks were shot by hunters during regular hunting season and presented to wildlife officers." Thus the disease, having escaped from the wildlife research facilities, began spreading quickly among elk and deer.

Relentlessly, throughout the 1980s and the 1990s, CWD spread outward from the original infections surrounding the Colorado State University facility. There were two main vectors that accelerated this spread. One was the normal traffic and migration pattern of deer and elk in and around the state of Colorado. But by far the fastest way of spreading CWD was via the burgeoning billion-dollar industry of farmed cervids (deer and elk). In the latter part of the twentieth century, elk and deer farming took off exponentially. These animals were bred in captivity and used primarily for hunting on private preserves and for venison and antler sales. The transport of elk and deer within states and between states became routine in the 1980s.

Not only was the disease spreading within the United

States, but it also spread across the border. In April 1996, the Southeastern Cooperative Wildlife Disease Study (SCWDS) in Canada released the following statement: "Chronic Wasting Disease was recently diagnosed in a privately owned elk in Saskatchewan, Canada. The initial diagnosis was made by Saskatchewan Provincial Veterinary Laboratory and confirmation testing was performed at the USDA's National Veterinary Service Laboratories in Ames, Iowa. The affected cow elk was exported from South Dakota as a calf in December 1989. It was thought that its dam came from Colorado, but the history is vague at present. During the next 6 years the elk was at three different premises in Saskatchewan."

The announcement was worrying for several reasons. Not only had CWD spread internationally, but during the lengthy seven-year period from 1989 until the animal's diagnosis, the elk had been hauled through at least three properties in Saskatchewan. The announcement sent shock waves through the approximately 2,500 cervid producers in Canada, among which are about 1,250 elk ranchers. One of the more lucrative parts of Canadian elk farming was the export of velvet elk antlers to South Korea, China, and other Southeast Asian countries where they have been used in medicinal and nutritional traditional supplements for 2,000 years. Any threat to this rapidly increasing export business meant economic hardship for hundreds of Canadian elk farmers.

By the late 1990s, CWD had spread through Wyoming and had entered Nebraska and South Dakota. The spread through South Dakota is an example of the relentless march of the prions. A state government information release describes this almost military advance: "In South Dakota, CWD was discovered in six private, captive elk herds during the winter of 1997–98 and in another private, captive elk herd in August of 2002. CWD was first found in free-roaming wildlife in a white-tailed deer in Fall River County during the 2001 big game hunting season. Surveillance of free-roaming elk and deer by Game, Fish and Parks (GFP) has detected twelve infected deer and two infected elk. Wind Cave National Park has detected three elk and five deer that were infected with CWD."

The western panhandle of Nebraska abuts northeastern Colorado and it is here that the majority of cases occurred. Sioux and Kimball Counties have been the hardest hit in Nebraska. The pattern of concentration of CWD cases in game farms followed by their spread into the wildlife population is perfectly illustrated by a Nebraska state release: "Upon learning that a number of captive whitetail taken inside the Sioux County game ranch tested CWD positive, and concerned about the Kimball County results, Commission staff in January 2002 began a culling operation within a 15-mile radius of the Sioux County game ranch. Of 113 wild animals taken in that culling operation, nine tested positive for the disease, for an overall infection rate of nearly eight percent. Of those testing positive, five were culled within two miles of the game ranch boundaries, two were culled within two to five miles, and two were culled within five to seven miles. At the same time, Commission staff culled 172 mule and white-tailed deer from within the captive game ranch in Sioux County. Of 154 test results received, 79 animals tested positive."

CWD in captive herds can reach astonishingly high

prevalence. In addition to the original epidemic that killed over 90 percent of the mule deer at Colorado State University in the 1970s, a greater than 50 percent prevalence rate has been found on a Nebraska elk farm and prevalence rates of between 59 and 71 percent have been recorded in other captive herds, according to Williams. Like all other prion diseases, CWD has no cure.

The mode of transmission of CWD is different from that of BSE, however. It appears to be more like scrapie. In the United Kingdom, BSE seemed to have been transmitted primarily through the contaminated feed system (although there are now nearly fifty cases in the United Kingdom of BSE-positive animals born after 1996, when all of the feed loopholes should have been closed). CWD seems to transmit from animal to animal and there are indications of vertical transfer from mother to offspring. More worrying, as with scrapie, a pasture that is contaminated by CWD can remain contagious for several years. Drs. Williams and Spraker at Wyoming have documented captive cervids coming down with CWD when put to graze on a pasture that had been unoccupied for several years. It is extremely difficult to decontaminate a farm or a pasture from CWD infection. The apparent ease with which CWD spreads horizontally from animal to animal also means that as CWD amplifies within a given captive herd of elk or deer, unless stringent precautions are taken involving double layers of fencing around the property, it is extremely difficult to prevent the disease from spreading from captive out to wild animals in the area. Thus, every infected captive farm becomes a hot spot in the environment from which the disease can rapidly spread into the wild cervid population. Deer and elk migration patterns then efficiently spread it farther afield.

The spread of chronic wasting disease from its epicenter in Colorado into Nebraska, Wyoming, and South Dakota is relatively easy to explain. Infected deer simply migrate. Once CWD had reached the western slopes of the Rocky Mountains in Colorado, it decimated the state's much vaunted hunting industry. And it was only a matter of time before it reached Utah.

On February 18, 2003, a mature buck deer taken by a hunter on Diamond Mountain, just north of Vernal, Utah, tested positive. By the end of 2003, Utah had eight positive cases. Wildlife officials I spoke with told me this pattern was predictable and was factored into the planning. Not factored in were two sudden and unexpected leaps of CWD in 2002 and 2003. Out of nowhere CWD suddenly appeared in Wisconsin and in southern New Mexico. I interviewed some wildlife experts who professed puzzlement and alarm at the sudden appearance of the disease far away from its epicenter.

The first case of CWD in New Mexico was announced in June 2002, followed by three more cases in February 2003. What was shocking to wildlife biologists was that the deer came from the White Sands area at the southern end of the state, far south of the known epicenter of the CWD epidemic, located primarily in northeastern Colorado, Nebraska, and Wyoming. By February 2004, a total of seven mule deer had tested positive, all around White Sands. The finding of CWD there raises questions about how long the CWD has been in New Mexico wildlife.

I interviewed several CWD experts, as well as game and

wildlife officials from the state of New Mexico, about the unexplained emergence of CWD near the White Sands missile range. All experts pronounced themselves mystified about how an outbreak could have occurred at such a great distance from the epidemic's known epicenter. All the usual possibilities (migration patterns, interstate transport of animals) had been examined and eliminated. Is it possible that the CWD had migrated north from Mexico? Or alternatively, is it possible that CWD arose "spontaneously" in New Mexico? Or is it possible that deer in White Sands got CWD from previously infected wildlife in New Mexico?

Even more troubling, however, was the appearance of CWD in faraway Wisconsin. The first case of CWD in Wisconsin occurred on February 28, 2002, in three deer just west of Madison. The announcement created waves, because until then the disease had mostly been confined to the Colorado, Nebraska, and Wyoming epicenter. CWD had apparently jumped across the states of Iowa and Minnesota to land in Wisconsin. This jump was deeply troubling to wildlife officials who saw a high level of urgency in tracking down the mechanism of spread. "I can't even describe how I felt," Bill Mytton, the Department of Natural Resources' now retired big-game specialist, told *Milwaukee Magazine*. "It was this sickening fear." Hunting is a more than $1 billion business in Wisconsin and there was a sense of panic in the wildlife community. A major public relations offensive was called for.

James Kazmierczak, a Wisconsin Department of Health and Family Services epidemiologist, led the charge. "You could live on a diet of deer brains and never get sick," he said in an interview with *Milwaukee Magazine*, published on September 13, 2002. "There is either no, or very low, potential for infection in humans." The remarkable interview was conducted as part of an investigation into the growing evidence of widespread CWD infections in Wisconsin deer and in an atmosphere of increasing alarm on the part of hunters. Going into the fall 2002 hunt, Wisconsin state officials estimated there were 1.6 million wild deer, but the number that had tested positive for CWD, as of the end of 2003, was 210.

Kazmierczak's statement became famous nationwide. Wisconsin, of course, is well known for its outbreaks of transmissible mink encephalopathy (TME). Four out of five known TME outbreaks since 1947 have occurred in Wisconsin. The most recent outbreak occurred in Stetsonville in 1985 and lasted five months. Thousands of animals died. This, combined with the huge density of wild deer in the state, prompted Allen Boynton, wildlife biology manager for the Virginia Department of Game and Inland Fisheries, to tell *Milwaukee Magazine*: "If Wisconsin doesn't stop CWD, it will move all the way across the country. This will become a national story."

In 1995 alone, the records of the Wisconsin Department of Natural Resources show, 26,488 road-killed Wisconsin deer were rendered into feed. This feed was used to feed other ruminants, notably cattle, but also a wide variety of other farm animals and even some captive cervids (deer and elk). Since there was nothing remarkable about 1995, this figure can be taken as an annual road-kill average for Wisconsin. If any of these 26,488 animals was infected with CWD, then CWD was likely passed on via the rendering process to deer and perhaps cattle. And lest there be doubt that CWD can transmit to cattle experimentally, a recently published study from the USDA labs at Ames, Iowa, has unequivocally demonstrated transmission of the CWD agent to cattle via brain injections. Fortunately, the rules regarding roadkill and feed have been tightened over the last decade, though they are not yet airtight.

More bad news hit on March 11, 2004. The Associated Press reported that a new test procedure for CWD called IDEXX had turned up a large increase in cases in Wisconsin. The test revealed that in addition to the eight counties where the disease had been found in deer originally, deer had tested positive for the disease in fourteen additional counties, some of which were more than 200 miles away from the original hot zone. The spread of the disease was to have a severe impact on the $1.6 billion hunting industry.

Many greeted this latest report with disbelief, saying the test method was new, unproven, and inaccurate. Follow-up news stories quoted officials who claimed that the IDEXX tests were not intended to be 100 percent accurate and that subsequent confirmatory tests, using the "gold standard" IHC antibody test, showed only forty-five new positive cases. But as anyone who has worked in an immunology lab knows, an antibody-based test is only as good as the antibody and the pathologist. Both IDEXX and IHC tests are antibody tests and both therefore are subject to some level of error. Though IHC is probably more accurate than IDEXX, calling it the "gold standard" may be overstatement.

In January 2004 more CWD-positive deer were found in Kenosha County, considerably farther east than the traditional "killing zone" in Wisconsin, where over 200 deer have already tested positive for CWD. "Very clearly, there needs to be a lot more testing on cattle in the U.S.," University of Wisconsin–Madison professor Judd Aiken told the Associated Press. "Will this divert some of the efforts toward deer disease? I guess that is possible." Aiken's concern about future testing capabilities came as CWD moved farther eastward in Wisconsin with state officials confirming that a deer in Kenosha County tested positive for the illness. Six Wisconsin deer tested positive for the disease in Kenosha, Rock, and Walworth counties in January 2004. And Illinois has identified 30 diseased deer in their border counties, according to wildlife officials. There is widespread agreement that it is only a matter of time (and testing) before CWD is picked up in Iowa. When that happens, CWD will be epidemic in a band that stretches from Utah all the way across the Mississippi to Milwaukee.

Two questions haunt wildlife officials in many states these days. The first is: Can CWD spread to cattle? During the fall and winter of 1997 and 1998 Colorado State University and other institutions launched an official monitoring program of cattle in the area of the CWD spread. Only cattle more than four years old that had been on the property on which deer had been grazing in a CWD hot spot more than four years were examined. They published their results in May 2003: "Findings from this study suggest that large-scale spread of CWD from deer to cattle under natural range conditions in CWD-endemic areas of northeast Colorado is unlikely."

This conclusion, however, is questionable for several reasons. Though it's true that the cows were grazing in the same "game management units" (GMU) as deer or elk that had been diagnosed with CWD, these GMUs sometimes covered several square miles and there is only a very small probability that the deer and cattle crossed paths in such a large area. In fact, the study specifically alludes to this low probability. A few score old cattle were sampled from only seven game management units, but the probability of actually obtaining a cow that had been in close contact with a CWD-infected deer under these conditions was very low. The only true test of whether CWD can spread from deer to cattle would be to *randomly* test a much greater number of cattle in each GMU. This was not done.

While the primary reagent used in this study is considered a "gold standard" for diagnosing pre-CWD infection, whether it is also capable of diagnosing a prion strain induced by jumping from deer to cows is not known. Certainly, in the circumstances of cattle grazing with deer, a preclinical prion infection in the cattle would have been very difficult to detect pathologically. Even a negative result with this antibody test does not completely rule out prion infection in these cattle. And to top it all off, the researchers *did* find some neuropathologic changes in some cattle, which they dismissed with the conclusory statement that "Some incidental neuropathologic changes unrelated to those of TSEs were detected."

The other question that haunts wildlife officials these days is: Can eating CWD-tainted meat cause disease in humans?

18

More Tainted Meat

The McEwans were a young married couple living in Syracuse, just north of Salt Lake City, Utah, in 1998. Doug McEwan was an enthusiastic and frequent hunter and he relished eating the tasty venison from deer and elk kills. He was twenty-nine when his wife, Tracie, first began to notice that his memory was not what it used to be.

"My husband began having problems that I noticed in early summer," wrote Tracie in a letter for a support group that was published in the *Rocky Mountain News* in 2002. "The first thing I remember was he forgot how to spell my name. Then I just noticed things like if I called him at work and asked him to bring home milk, he would forget the milk or that I had even called. He was having problems getting all of his paperwork done, so I was doing his monthly expense

report. I didn't realize until about a month later that he was
actually having a hard time doing basic math. By the end of
July, he was really having trouble. He was in Idaho on business, and was very late calling home one night. When he did
he told me he couldn't remember our phone number."

Doug's memory problems worsened over the summer and
within a couple of months he resigned from his busy job
because he could no longer function. After multiple uninfor-
mative tests, Doug had a brain biopsy in November. The
results were obvious; Doug had CJD. The doctors were puzzled: CJD at age thirty? The average age for CJD was in the
sixties. Doug's brain lesions did not match the lesions
described for variant CJD, the human variant of mad cow disease that had been seen in the United Kingdom. The doctors
assured Tracie that this was a random occurrence and government specialists who examined Doug's condition concurred.
Doug died a few months later from CJD. After reading about
mad cow disease and CWD, Tracie was convinced that
Doug's passion for hunting and consumption of venison
might have had something to do with his untimely and
bizarre death. But officially the Centers for Disease Control
determined that there was no strong evidence that Doug
McEwan's death was connected to CWD.

In 2003 I interviewed a prominent and high-profile epidemiologist who had investigated the deaths of three young
people who had consumed venison. The epidemiologist, who
did not want to be identified in print, informed me that there
simply was not sufficient evidence to draw a cause-and-effect
relationship between CWD and the untimely deaths of these
young venison consumers, two of whom were hunters. He

went on to recommend an in-depth epidemiological study of the hunters in northeastern Colorado who had consumed venison from this known CWD-endemic area, but the epidemiologist lamented the lack of political will in the Colorado governor's office to launch such a study, mainly because of "privacy concerns." Hunters come to Colorado from all over the United States, so a systematic tracking of hunters' names and addresses as well as their health status over time would be necessary. This project would take years because of the lengthy incubation period of prion diseases.

The governor's office in Colorado is unlikely to rock the boat anytime soon by introducing such a controversial measure. It would draw attention to Colorado as the epicenter and likely origin of the CWD epidemic and it would cost the state an enormous amount of money. Since sporadic CJD is still not a reportable disease in most states, and because nobody has succeeded in firing up momentum in Colorado to conduct the definitive epidemiological study, there are no strong data to either support or negate a direct link between CJD occurrence in humans and the eating of CWD-infected venison. However, it could be that there is no link because nobody has looked carefully enough to find it.

The fact that thousands of hunters from Colorado and Wyoming have *not* come down with CJD is routinely cited by state officials and by the Centers for Disease Control personnel to reassure the public that CWD does not pass to humans, in spite of the fact that no systematic study has ever been done. It is this kind of "hall of mirrors" circular argument that brings a sense of déjà vu when the reaction of the British government in the 1990s to the BSE epidemic is compared with U.S. authorities reacting to the CWD and possible BSE cases in the United States.

During the 1980s the legendary Carleton Gajdusek investigated the possibility that CJD might be a "zoonosis," an animal-derived disease. The paper he published on this topic with veterinarian researcher Zoreh Davanipour has been generally overlooked, but what Gajdusek found cannot be easily dismissed. There was a statistically significant linkage between CJD and deer in those who both worked with deer or in those who were associated with deer as a hobby (hunting). Gajdusek's team also found a curious linkage between CJD and people who worked with monkeys and squirrels and who hunted rabbits. Could this mean that in the United States prion diseases are running quietly through wild animals, being spread among species? While there have been no follow-up studies to Gajdusek's work, his data are certainly arresting and they point in the direction that wildlife in general, but deer in particular, may be widespread carriers of prion diseases.

Anyone attempting to validate this wildlife infection theory will quickly stumble upon a bizarre cluster of CJD deaths in Kentucky that involved a group of people who ate squirrel brains. The study was led by Dr. Joseph Berger from the University of Kentucky and was reported in *The Lancet* in 1997. The three men and two women, aged from fifty-six to seventy-eight, came from a rural part of Kentucky where people traditionally scrambled squirrel brains with eggs. On other occasions they cooked the squirrel brains in a meat and vegetable stew called "burgoo." The deaths of five people who ate wild squirrels when seen in the light of Dr. Gajdusek's study has worrying implications that go beyond the CWD epidemic. Especially as the deaths in Kentucky were not the only ones linked with eating squirrel brains. A previous study in 1984 documented the deaths from CJD of people from East Texas and from Louisiana who had also eaten squirrel brains. The data appear to imply that a form of prion disease may be endemic in wildlife in some parts of the United States.

In 2002, the Centers for Disease Control investigated another report of three men who had died from a brain disorder, allegedly as a result of eating wild game feasts. Though only one of the three had actually died from CJD, the report noted: "Although no association between CWD and CJD was found, continued surveillance of both diseases remains important to assess the possible risk for CWD transmission to humans." Without a more comprehensive and systematic way of tracking the deaths of hunters from CJD, it will remain difficult to support or negate the increasingly disquieting hypothesis that CWD can cause CJD.

But a closer look at some people who may have eaten CWD-tainted venison and later developed CJD indicates that the notion of a link between the two is not fantasy. Dr. Ermias Belay from the Centers for Disease Control examined the case histories of three unusually young patients who had died of CJD. Patient #1 was a twenty-eight-year-old woman who died in June 1997 after a four-month illness and autopsy showed she had CJD. The patient had consumed deer meat as a child. The deer were harvested in Maine and New Jersey. When she was six, on two different occasions she ate venison from a deer likely harvested in Wyoming. Patient #2 died at age thirty in March 1999. The patient was a regular hunter and had hunted almost every year since 1985. Patient #2 hunted in several areas, mostly in Utah, but also in southwestern Wyoming in 1995 and in British Columbia, Canada, in 1989. Additionally, the family on several occasions received meat from the patient's brother, who hunted in Utah. The patient regularly ate liver from deer and elk, but no other internal organs. Patient #3 was a twenty-seven-year-old truck driver, began experiencing symptoms in December 1998, and after a fifteen-month illness died of CJD in April 2000. Patient #3 was described as an avid hunter and had hunted since age thirteen. Although there was no record of patient #3 eating deer from Colorado or Wyoming, the custom processing plant that processed his deer also processed about twenty elk from Colorado every year. In other words, elk from a potentially CWD hot zone were processed by the same plant.

In arguing that these three cases did not provide strong evidence for a cause-and-effect relationship between eating tainted venison and early-onset CJD, the Centers for Disease Control scientists cite supporting evidence that 299 deer brain samples from patient #1's eating zone, 404 deer and elk samples from patient #2's eating zone, and 138 samples from patient #3's eating zone were submitted to the National Veterinary Services Laboratories, USDA, at Ames, Iowa, and all were found negative for CWD using the IHC test. This allowed the scientists to state that the link between the young people eating venison and dying of CJD is weak.

But there are serious problems with this argument. When the study was published in 2001, no CWD cases had been reported in Utah, where patient #2 had done most of his hunting. But beginning in February 2003, eight positive CWD cases had been reported from around Vernal in northeastern Utah and Moab in east-central Utah. This new information, which was unavailable to the Centers for Disease Control when they published their paper, strengthens the epidemiologic link between the death of patient #2 and his hunting habits. In addition, it raises the question: If there is no definitive link between eating venison and deer testing positive for CWD, does that exclude linking the two? The answer is no. One feature of the rapid spread of CWD is that experts have been constantly surprised by the unexpected twists and turns of this mad deer epidemic. For example, nobody expected CWD to suddenly appear out of nowhere and with such sudden virulence in Wisconsin. Second, the accuracy and integrity of the CWD (and BSE) testing data coming out of the Ames, Iowa, labs has recently come under severe criticism.

USDA veterinarians have openly said that the results coming out of Ames, Iowa, cannot be trusted and often conflict with tests from other labs. In other words, there is extreme skepticism about the competence and integrity of the testing process at Ames, Iowa. Department of Agriculture veterinarians and a deer rancher told United Press International in February 2004 that "[t]he federal laboratory in Ames, Iowa, that conducts all of the nation's tests for mad cow disease has a history of producing ambiguous and conflicting results—to the point where many federal meat inspectors have lost confidence in it. . . ." The tests with negative results, performed by a lab whose credibility has been called into question, in deer from the areas around which the three CJD victims ate venison comprise the bulwark upon which the arguments by the Centers for Disease Control rest.

Finally, the Centers for Disease Control admit that attempts to weaken the link between CWD and CJD are not helped by the lack of accurate reporting of CJD cases. The report states that "During 1979–1996, only 12 CJD cases in this young age group were reported to the CDC." But it is an open secret that many, if not most, cases of CJD are not reported. As stated earlier, the national prion surveillance system is woefully underfunded; it relies heavily on the stalwart efforts of a group of volunteers, many of whom have had family members die of CJD, to collect the statistics for CJD cases. Therefore, any citations of CJD statistics in young people are almost certainly underrepresented.

At the annual meeting of the American Academy of Neurology in Denver, Colorado, at the end of 2002, Dr. Norman Foster's group from the University of Michigan presented some alarming new data. Two young men, aged twenty-six and twenty-eight, had shown up at the Neurology Institute of the University of Michigan at the same time suffering the same symptoms. Although no link to eating venison had been found at the time of publication, the researchers declared, "The concurrent development of CJD in two unrelated young men in a single state suggests that sporadic CJD may be more prevalent than previously realized."

There is an eerie, déjà vu quality to the conclusions coming out of the Centers for Disease Control and the USDA regarding the possible link between eating venison and CJD.

It is reminiscent of the British government's stumbling attempts to "spin" the increasing amount of evidence of a link between eating BSE-tainted beef and the emergence of variant CJD in young people prior to the March 8, 1996, Damascus experience when the data suddenly became irrefutable. The old often-heard Centers for Disease Control argument is that since these patients in the United States do not have variant CJD (human mad cow disease), then we can go back to saying that the three young venison eaters died from an unknown but random cause. But, of course, since CWD is a relatively new disease in deer and elk, nobody has any idea of what to expect regarding what form of CJD these people should die of. The distinction between so-called sporadic CJD and variant CJD in humans has received much press in the United States, but recent scientific data have shown this distinction may be little more than a comforting mirage. That there is *any* possibility of a link at all between eating venison and CJD should be raising a bright red flag.

When the Europeans, who have just come through a gut-wrenching ten years of the BSE epidemic, resulting in millions of animals slaughtered and at least 150 human fatalities, look across the Atlantic and see the fumbling attempts to deal with CWD on this side of the Atlantic, many shake their heads. They have seen this before. Speaking at the Days of Molecular Medicine conference in La Jolla, California, in March 2002, European prion expert Adriano Aguzzi issued a strong warning against underestimating the dangers of CWD in the United States. "For more than a decade, the U.S. has by-and-large considered mad cows to be an exquisitely European problem. The perceived need to protect U.S. citizens from this alien threat has even prompted the deferral of blood donors from Europe," he said, according to the journal *Nature Medicine*. "Yet, the threat-from-within posed by CWD needs careful consideration, since the evidence that CWD is less dangerous to humans than BSE is less-than-complete."

Aguzzi stated that CWD is probably the most mysterious of all prion diseases. "Its horizontal spread among the wild population is exceedingly efficient, and appears to have reached a prevalence unprecedented even by BSE in the United Kingdom at its peak. The pathogenesis of CWD, therefore, deserves a vigorous research effort. Europeans also need to think about this problem, and it would be timely and appropriate to increase CWD surveillance in Europe too."

No plan of action on the CWD problem has yet been implemented in the United States. Even though the warning signs are there, action plans remain on various shelves, gathering dust.

19

The Monitors

The spread of prion disease in the United States seems to have caught everyone by surprise—but it may not have gone completely unnoticed. Someone appears to have been monitoring the possible risk that prions may pose to the nation's food supply for the past four decades. While this statement is speculative, there is considerable circumstantial evidence to back it up.

A few years after veterinarian Gaylord Hartsough presented strong scientific data before researchers at a National Institutes of Health conference in 1964 that cows could be harboring a new scrapielike disease, a mysterious and chilling phenomenon arose in the western part of the United States. Ranchers and cattle owners began to find their cattle dead, not from disease, or lightning strikes, or from other, familiar natural causes. They didn't know who or what had killed their cattle, but one thing was clear—somebody had selectively removed their organs. In some cases, cuts left behind suggested that sharp knives or scalpels had been used. It looked as if the organs had been surgically removed. The most frequently removed organs were the eye, the tongue, the reproductive organs, and the anus or large intestine. Most disturbing of all was the fact that the animals were usually killed in the dead of night.

This "dead cow with surgery" phenomenon spread quickly through the United States beginning in the late 1960s, then accelerated through the 1970s, and has continued through the 1980s, 1990s, and into the present century. From the beginning, ranchers, and particularly the press, called these gruesome killings "cattle mutilations." Hundreds, if not thousands, of animal mutilation reports were investigated by local law enforcement, with cases occurring in fifteen states, from South Dakota and Montana to New Mexico and Texas.

By the mid-1970s, for example, scores of cases had occurred in northeastern Colorado, a major cattle-producing area. The cattle mutilations had begun to appear in the area around Brush, Fort Morgan, Greeley, and Sterling, all in the northeastern part of Colorado where prion disease in deer had just been discovered. In fact, there is a very interesting overlap in the spread of CWD in Colorado (and Wyoming) and the increased number of reports of cattle dying mysteriously with surgical cuts and the removal of organs that subsequent research has shown are loaded with prions.

In Colorado, as elsewhere, the modus operandi of the cattle surgeons was always the same. The previously healthy animals were found lying on the ground with an eye (usually the left eye), an ear, the tongue, sex organs, or anus and large intestine removed in clean, bloodless incisions. No tracks were apparent, either from humans or vehicles in the vicinity of the dead animals. In a two-year period from 1975 to 1977 in two Colorado counties alone, there would be nearly 200 reports of mutilated cattle.

On September 4, 1975, Governor Richard D. Lamm flew to Pueblo, Colorado, to confer with the executive board of the Cattlemen's Association about the mutilations, which he called "one of the greatest outrages in the history of the western cattle industry." The governor noted, "It is no longer possible to blame predators for the mutilations."

The press now began to take notice of the mutilations. The article that appeared in the September 22, 1975, issue of the *Glenwood Springs Post*, published in Glenwood Springs, was one of hundreds that appeared in Colorado and throughout the Midwest in the 1970s, according to Frederick Smith's book *Cattle Mutilations*. "Garfield County Undersheriff Hart told the *Post* last week that Garfield County was the scene of cattle mutilations as early as February of 1975," the article stated. "Hart also reported that on September 13, a Hereford cow was discovered mutilated just across the Garfield county line in Mesa County. Missing were the animal's sex organs, rectum and an ear."

A couple of months later, the language reporters used to cover the story had escalated—as had the rewards. "Rewards now total $23,000 for information leading to the arrest and conviction of stock mutilators in Colorado," stated an article in the December 1975 issue of the *Farm Journal*, a national publication. "The last $10,000 was posted by the Elbert and

El Paso County Ranchers and Farmers Association for incidents in those two counties. More mutilations have been reported in New Mexico, Nevada, Utah, Oregon, Wisconsin and Iowa. Shots have been fired at Army helicopters. . . . But lawmen are still looking for the first solid clue."

It should be noted that Wisconsin and Minnesota were among the first states to be hit with cattle mutilations back in the early 1970s. On October 13, 1974, a buck deer was found in a field eighty-five miles south of Twin Cities, Minnesota. Sheriff Elroy Johnson of Brown County reported that the deer was found dead of unknown causes and its sex organs removed. A veterinarian by the name of Calvin Glenn examined the animal and concluded the sex organs had been removed with a knife. Side by side with this curious killing of a deer was an epidemic of cattle mutilations. Between 1970 and 1974 one source reported that twenty-two mutilated cattle had been reported in Minnesota. Wisconsin was also hit hard in the early 1970s. About thirty years before the advent of officially reported CWD in deer in Minnesota and Wisconsin, a veritable epidemic of mutilations occurred in both cattle and deer. No one is even thinking of a link between cattle and deer mutilations and CWD, of course. That's beyond the pale.

In any case, by the beginning of the 1980s cattle mutilations had become a law enforcement enigma. Nobody had been caught or charged in the hundreds of serial killings in the western United States. Police in several states launched intense investigations into the mysterious deaths and were almost universally unsuccessful in apprehending the perpetrators. Mostly because of media hype and those claiming that UFO aliens were conducting the mutilations, the phenomenon became a laughingstock. This guilt by association ensured that no scientist or veterinarian would take the issue seriously. Whenever the phrase "cattle mutilations" was mentioned, most intelligent people rolled their eyes and quickly excused themselves from the conversation. But because of the numbers of cattle dying mysteriously and probably illegally, the law enforcement community had no choice but to take it seriously.

Several intensive investigations by highly professional lawmen in several states turned up evidence that large numbers of cattle were being killed and that their organs were being removed with sharp instruments. I spoke with retired sheriff Tex Graves, who used to be sheriff of Logan County, Colorado. Sheriff Graves investigated more than a hundred cattle mutilations in the Greeley, Sterling, and Fort Morgan area. There were always strange, silent aircraft in the vicinity before and during the mutilations, he told me. Sometimes what looked like silent helicopters were reported flying in the area.

Graves had accumulated a library of photographs of mutilated cattle over the years. He told me he had to restrain the local ranchers from shooting at, and bringing down, the mysterious helicopters and aircraft. The local newspapers carried stories about the mutilations almost daily in the 1970s together with graphic photographs of dead cows with their sex organs and anuses or large intestines removed. Sometimes the newspapers carried a quote from a veterinarian saying that the most likely cause of cattle mutilations was predators and scavengers. Two decades later, "predators" was still the favored official explanation.

•   •   •

It was in the 1990s that my decade-long scientific interest in prion diseases turned into active research into an area of study that I, at first, thought had nothing to do with the spreading prion epidemic. In the previous decade, I had obtained my Ph.D. in biochemistry and begun my career in cancer research. I continued my research career in molecular genetics and immunology both in Canada and in the United States, and then in 1996 I joined an organization in Nevada called the National Institute for Discovery Science whose mission was to apply the scientific method to study alleged anomalies. During this time I became project manager for conducting forensic research into the rash of unexplained cattle deaths that were occurring in the western part of the United States. The project involved the biochemical forensic analysis of tissue samples from cattle that had died of unexplained causes in an effort to determine cause of death.

It was in the course of this work that in 1998 a colleague and I paid a visit to one of the most prestigious veterinary research labs in the country, Colorado State University, which is located in northern Colorado. For half a day we sat around a meeting room with a group of veterinary specialists. Dr. Barbara Powers, the head of the veterinary diagnostic lab, assured us that they had conducted a study on cattle mutilations and that their consensus was that they had never seen a case of "cattle mutilations" that they could not explain by simple predation. This sort of dismissal-by-predation theory provided the route by which the mysterious killings were ignored by the dairy and cattle industries.

But a close look at the phenomenon, without any preconceived notions, reveals a far more disturbing truth. During my eight years tracking these mysterious cattle surgeries, I learned to listen carefully to what the ranchers, local veterinarians, and law enforcement told me, but without necessarily believing everything I was told. I felt that a healthy skepticism, as opposed to debunking, was an appropriate stance. One thing that wildlife experts would frequently point out to me was the uncanny resemblance between the pattern of organ removals that were taking place in cattle mutilations and standard wildlife sampling techniques for monitoring the spread of infectious agents in the wild.

The first of these parallels involves the use of helicopters. Dozens, if not hundreds, of anecdotal reports from ranchers and police claim the presence of black or unmarked helicopters in the area of cattle mutilations. One study of a particularly well-documented cattle mutilation wave in the vicinity of Malmstrom Air Force Base in Montana, which was investigated by the Cascade County sheriff's department, showed a statistically significant link between the time and locations of animal mutilations and the presence of helicopters and unidentified aircraft.

In wildlife research, some of the most successful stalking of free-ranging animals is carried out from helicopters, either during the day or at night with vehicles equipped with spotlights. In 1993, for example, when anthrax was causing significant mortality in wildlife west of the Great Slave Lake in Canada's Northwest Territories, researchers located 55 percent of the carcasses visually using remote infrared sensing cameras mounted externally on a helicopter. It makes a lot of sense to use helicopters when tracking an infectious organism in thinly populated rural areas. On occasion, field biologists have used the downdraft from helicopter rotors to flush out animals from brushy hiding places, and helicopter maneuverability is exploited effectively to get gunners within dart range of selected animals.

Exactly the same problem applies to tracking the spread of prions through the nation's cattle. Again, from the perspective of sampling, a lot of ranchers told me that they thought their animals had been lifted up into aircraft, mutilated there, and dropped back onto the ground. In one case in rural Montana, a local veterinarian told me he was called out to conduct a necropsy on an animal that lay in fresh deep snow. When he went out to the animal there was only a single pair of tracks in the snow to and from the dead animal— and they were from the rancher who had discovered the dead cow. The veterinarian told me the cow had been almost completely exsanguinated (all blood removed). It lay in a depression in the snow and had probably been dropped from the air onto the ground after the mutilation. Obviously an aircraft of some type had been used.

I came across this dropped-from-the-air aspect of cattle mutilations repeatedly. Another rancher I spoke to told me that one of his mutilated cows had four broken legs and had been dropped from the air onto the ground. And in a series of investigations in northern New Mexico in 1978–1979, highway patrol officer Gabe Valdez documented the presence of abrasions and cuts around the hocks of some mutilated cows. Valdez told me they were reminiscent of some clamplike device that might have been used to secure the animal and lift it to another location by means of a helicopter.

In 1998, Valdez and I investigated a case of cattle mutilation just outside Los Brazos, New Mexico. The animal, a four-year-old Red Hereford, was lying about a mile from Highway 64 and within a few hundred yards of a lake. Our examination indicated that the animal had been dead about forty-eight hours and that the left eye, tongue, and reproductive organs were missing. We noticed that the rest of the herd kept its distance from the animal as we arrived, and when we entered the pasture, a few cows even bolted in the opposite direction. The herd was plainly nervous. As always, we split up and spent several hours quartering the ground in a three-hundred-yard radius of the dead cow. This is standard police crime scene procedure.

After two hours of careful scrutiny of the ground, a pattern emerged. A trail of upturned cow pies could be seen emanating for several hundred yards from the location of the dead animal and heading east. Valdez's experienced eye had spotted the pattern. As I walked slowly east of the animal following the trail, the significance of this evidence dawned on me. The downdraft from a low-flying helicopter rotor would perfectly explain this effect. The copter would have flown in from the east and perhaps darted the cow as it arrived. The surgeons could then have dismounted from the craft and within a few minutes removed the organs from the animal before departing in the helicopter again to the east. At the time, the ground was too dry and hard to reveal the presence of footprints.

The second parallel between mysterious cattle surgeries and the methods used by wildlife biologists to track the spread of infectious disease involves the use of a chemical-like formaldehyde. Witnesses tell us that predators and scavengers sometimes avoid mutilated carcasses. This avoidance of the mutilated carcass by scavengers has assumed legendary proportions in the lore of cattle mutilations. Ranchers have told me they have seen coyote tracks sometimes circling a mutilated carcass at a range of thirty feet, but never coming in to sample the body. The avoidance may be the result of unnatural chemical odors lingering on the carcass.

In wildlife studies, such as the Canadian anthrax study, formaldehyde was applied to infected carcasses to prevent scavenging, which would further transmit the disease. Formaldehyde has a very pungent unpleasant odor; even traces of this chemical lingering on the body would be enough to discourage scavengers from venturing too close.

I think it is no coincidence that formaldehyde has been found on cattle mutilations. When we conducted a forensic investigation on an unusual blue gel found on a mutilated cow in northeastern Utah in 1998, a chemical analysis found low levels of formaldehyde in the gel. The animal was found lying in a strange posture with front legs tucked up under her and legs splayed out behind. A beautifully precise half-inch diameter cut had been taken from around her left eye. A photograph of the cut around the eye, which was published in 1999, came as a shock to many people because the precision of the cut was obviously far beyond anything ever discovered in a cattle mutilation.

During my investigation of this case, a veterinarian and two seasoned investigators had examined the animal and found the curious blue gel-like substance smeared over the eye. In one of the most thorough cattle mutilation investigations in history, the gel was carefully taken from the animal's eye and subjected to a battery of chemical tests to determine its composition. And what did we find? Low levels of formaldehyde. Because formaldehyde evaporates easily, the initial formaldehyde concentrations in the gel were probably considerably higher than those reported. If the animal found in Utah was sampled for an infectious disease, it would be standard operating procedure for the mutilators to try to prevent local wildlife from eating the tissue and spreading the disease farther into the food chain.

The third and final parallel to the mysterious cattle surgeries involves tranquilizers and sedatives, which are routinely used to immobilize and euthanize wildlife prior to tissue sampling. The cattle mutilators know that they have time on their side and the longer an animal lies in the pasture before being discovered, the more likely decomposition will eliminate any traces of sedative from the animal. The personality of the rancher ensures the required time lag.

Ranchers are very independent. They have to be. In order to survive in the cattle business, they have to become expert problem solvers and troubleshooters. They are forced to rely on themselves and they rarely ask for help except in emergencies. In my experience, when a rancher finds a mutilated cow or bull, their first reaction is to say nothing. Generally, he or she will wait a day, maybe two days, before they pick up the telephone to call for help.

During the summer, when the heat and the humidity are high, that delay can seriously hinder an investigation. Because of this delay, I consulted a variety of forensic experts to determine what tissue would be most representative of what was going on in the animal at the time of death and what tissue would be slowest to decompose. In other words, I was looking for something that went beyond the standard frustrating litany of opening the animal and taking samples of decomposed blood and liver/kidneys, much of which was a smelly mush by the time we got to the animal anyway.

The answer I received from the forensic community surprised me. I was told that eye fluid is the slowest to decompose because it is less prone to bacterial infection. It also apparently provides a very good snapshot of what was happening at the time of the animal's death. So it was that we came to do some gas chromatography–mass spectrometry studies of the eye fluid of a small number of mutilations.

Our results showed much higher levels of oxindole in the eye fluid from the mutilated animal than in the eye fluid from a control (or sham-mutilated) animal. At low levels, oxindole is a metabolic by-product of tryptophan, an essential amino acid necessary for normal growth and development, but at high levels it is an extremely effective sedative in rats, humans, and dogs, causing decrease in blood pressure, loss of muscle tone, and the loss of consciousness. In other words, oxindole is a very intelligent compound to use if you want to hide your tracks, because depending on when the investigators get to the body, normal decomposition will begin to degrade oxindole into the kinds of substances that can be found in normal animals.

The same gas chromatography technique used in other mutilation cases uncovered a range of compounds, many of which could be metabolic breakdown products of sedatives. In a case from Valparaiso, Nebraska, and in another from Cache County, Utah, for instance, I found succinate in the

animal, which is a possible breakdown product of succinylcholine, a well-known sedative. Succinylcholine is a classic
sedative to use, because like oxindole, it blends unnoticeably
into the background. It was clear to me that a good deal of
intelligence was being applied to the selection of chemicals
used in these cattle mutilations. Whoever was doing it, they
were clearly not amateurs. They were skilled in the art of
deception and in covering their tracks.

Nor was it the first time that this sedative had been used
in a cattle surgery. Succinylcholine had also been found in
several cattle mutilations in Alberta, Canada, in the late
1970s. Dr. David Green, a veterinary pathologist from the
Alberta Provincial Veterinary Laboratory, had been working
closely with the Royal Canadian Mounted Police to investigate a wave of mutilations that began in the Calgary, Alberta,
area in 1979. Green necropsied twenty-four mutilated animals that year and another twenty in 1980. In a few cases he
had found needle marks and both succinylcholine and ketamine, a general anesthetic. In some cases, he determined that
the cows had died of ketamine overdoses. But it was the use
of succinylcholine that fascinated Green because it could be
so swiftly metabolized. "Succinylcholine is very hard to identify," Green noted in an interview with the authors of *Mute
Evidence*. "After it is metabolized it is broken down into substances that resemble the normal byproducts of body metabolism. It doesn't leave behind any telltale chemicals that are
easily identified."

Twenty years later we would find exactly the same sedatives in mutilated cattle. The perpetrators were following the
same basic modus operandi. Only rarely have they let down

their guard and failed to cover up their tracks. In one of those very rare cases where the mutilators made a mistake, Captain Keith Wolverton of the police department in Great Falls, Montana, was able to recover the evidence. It occurred when a metallic glint caught his eye as he and a veterinarian turned the animal over during the necropsy of a mutilated cow in a pasture twenty miles from Great Falls. Lying directly underneath the carcass was a large-gauge needle. There was really no other reason for a large-gauge needle to be lying under a mutilated cow in a remote pasture except if it had been used and dropped by mistake at the scene of the crime. For Wolverton, it was one of those rare "aha moments."

When he showed me the needle, I immediately recognized it as a standard issue, large-gauge needle of the type I had used for obtaining blood samples from the jugular veins of large animals if I wanted a very healthy flow of blood into my collection vessel. It would have been pretty easy to hook it into a catheter and a pump and to exsanguinate the animal in less than half an hour. Earlier, Wolverton and his team of veterinarians had found puncture marks in the jugular vein of a mutilated animal. The presence of medical hardware at the mutilation scene appears supportive of a sampling operation rather than cult activity.

When these parallels—use of helicopters, sedatives, and formaldehyde—between mysterious cattle surgeries and the methods used by wildlife biologists to track the spread of infectious disease are taken together with the timing of the phenomenon, cattle mutilations begin to look very much like acts of covert surveillance intended to monitor our food supply. The only rational purpose I could think of for such a monitoring operation would be to assess the level of prion infection in the nation's cattle. And the reason for the secrecy was obvious. Whoever was doing it wanted to avoid raising the alarm. But I didn't become totally convinced of the connection until I realized that the very organs taken in cattle mutilations are those that potentially contain high levels of prions.

Across the nation, local newspaper archives from 1975 to 1980 document thousands of animal mutilation reports. Though few methodical studies have been conducted on these reports, the anecdotal evidence suggests that the eye, ear, tongue, lips, reproductive organs, and anus are those most commonly reported missing in these animals. A survey conducted by the National Institute for Discovery Science would eventually confirm this anecdotal evidence. The survey was sent to 3,849 veterinarian bovine practitioners in the United States. Out of the 189 returned questionnaires, ninety-two reported mutilations, and the practitioners who replied to the survey reported thirty-nine cases in which the tongue had been removed, fifty-four cases documented the removal of an eye, and seventy each the removal of reproductive organs or rectum. So both anecdotal evidence and a nationwide survey of bovine practitioners indicate that the most common organs harvested in cattle mutilations are the reproductive organs, anus/large intestine, eye, and tongue.

How does this match up with the latest prion research? In the 1980s and 1990s, researchers attempted to determine where in the body—apart from the brain—abnormal prions concentrated. This is a crucial question if one is ever to discover the modes of transmission. For example, the standard explanation for the propagation of prions in cattle is through contaminated feed. When a cow ingests prion-contaminated feed the opportunity exists for the mouth, esophagus and stomach, small intestine, large intestine, and then internal organs to be contaminated with prions. All this occurs long before they begin to affect the brain—a phenomenon that occurs in cattle three to eight years later. Anyone tracking the prion spread in the nation's cattle herds in an intelligent manner would be concerned not only with the early indicators of prions in the body before symptoms appear, but also the infection of prions from mother to offspring. In order to address these two basic questions, the sites of initial infection (mouth, intestines) need to be sampled, as well as the reproductive organs for the passage to offspring.

The reproductive organs are important for another reason. For decades researchers have suspected that sheep shed prions during birth and that contaminated afterbirth, lying in the pasture, was one of the primary means by which a field becomes contaminated with prions. We know that sheep grazing on a contaminated pasture, even after it has lain fallow for three years, can contract scrapie. Now experimental data suggest that aberrant prions accumulate in regions of the placenta in sheep. Another study found the aberrant prions in certain parts of the uterus of pregnant ewes, although not in the developing fetus. Of course, scrapie in sheep has parallels in cattle, so reproductive organs are an obvious research target for the prion-sampling operation. Therefore, the removal of reproductive organs and the frequent sampling of fetuses is exactly what one would expect if cattle mutilations were in fact a research project aimed at identifying the rate of the spread of prions through cattle.

Scrapie is perhaps the most studied of the prion diseases since it has been around the longest. A recent study of natural scrapie infection of sheep showed a rapid accumulation of prions in the gut-associated lymphoid tissues during the early stages of scrapie infection. If one were looking for evidence of possible early stages of prion infection in cattle, one would cut out the anus and large intestine to take it back to the lab to test for prions. The "cored out" anus is perhaps the single most commonly reported feature of cattle mutilations.

The tongue, which is one of the favorite organs to be removed during cattle mutilations, may also contain high levels of prions. A groundbreaking paper published in the January 2003 issue of the *Journal of Virology* was big news. The study showed that the tongue in hamsters was a reservoir for very high levels of prions after initial infection.

Like the tongue, the eye is another favorite of cattle mutilators. Not surprisingly, perhaps, as prions are known to accumulate in the lens of the eye in humans infected with CJD. In fact, there have been several well-documented cases where corneal transplants have passed CJD from the corneal donor to the receiver. So prions are harbored in eye tissue as well.

There is obviously an uncanny coincidence in the most commonly removed body parts in cattle mutilations and the areas of the body where prions accumulate. Thus, if animal mutilations are a prion-sampling operation, the eye, the tongue, the anus/large intestine, and the reproductive organs would need to be removed for laboratory analysis. This is *exactly* what happens in cattle mutilations. This is not a smoking gun, but it provides strong circumstantial evidence that, beginning in the late 1960s, someone acutely concerned about the possibility of a scrapielike illness in cattle in the United States instituted a covert monitoring program of the nation's food supply. Just who was responsible for this covert monitoring operation is currently unknown.

20

Hot Zone

May 2003 was a bad month for Canada. On the 20th, the provincial government in Alberta issued a sobering news release that hit the media wires in overdrive and ricocheted around the world: "The Canadian Food Inspection Agency (CFIA) has quarantined an Alberta farm in an investigation of a single case of bovine spongiform encephalopathy (BSE), commonly known as mad cow disease. . . . Preliminary tests performed at a provincial laboratory and at the CFIA's National Centre for Foreign Animal Disease were unable to rule out BSE. The CFIA sent specimens to the World Reference Laboratory at Weybridge, United Kingdom, which has verified the presence of BSE. . . ."

"We remain confident in our beef and cattle industry," said Shirley McClellan, Deputy Premier and Minister of Agriculture, Food and Rural Development, "and we will support both the CFIA and our cattle industry in eliminating this disease from Canada."

From the moment the story hit the newspapers, the mantra that eating beef was safe began. Just two days later, on May 22, Canadian Prime Minister Jean Chretien sat down to a hearty meal of beef in an Ottawa restaurant to assure the anxious Canadian public that eating beef was perfectly safe. In doing so, he provoked widespread comparisons with the then British agriculture minister John Gummer, who posed for the cameras with his young daughter Cordelia while eating hamburger to assure the anxious British public that eating beef was safe. The photograph of Jean Chretien dutifully putting a large gob of beef into his mouth was a moment of supreme irony.

Marwyn and Lisa Peaster know a lot about just how great the distance between expectations and reality can be. The Peasters were shrimp farmers who left Louisiana for a better life in Canada. About two years after the Peasters moved to Alberta from Louisiana, they began to gather a herd of cattle. On August 23, 2002, Marwyn bought a group of cattle from a Saskatchewan farm. The owners were forced to sell because of the drought. Included in the group was the cow that Marwyn would later learn to his horror was infected with the brain-wasting disease. Within weeks of the announcement, the Peaster family cattle were gone, seized by the government, and all were destroyed and their brains tested for BSE. The family became the focus for vitriolic attacks and barbed comments from people across Canada. "It's not too bad," Lisa Peaster told the *Western Producer,* as she cradled their month-old baby, Nicolla, in her arms. "There's been some people make comments, but nothing you can't handle."

A letter to the *Edmonton Journal*, Alberta's flagship newspaper, summed up Canada's nationwide sense of frustration: ". . . [P]robably not since Mrs. O'Leary's infamous cow started the Great Chicago Fire of 1871—that killed 300 and made 100,000 more homeless—has a single cow done as much damage as Marwyn Peaster's lone BSE-infected cow has done to Alberta's economy in the past month. Estimates are that this week the impact on ranchers, feedlot operators, slaughterhouses, feed-grain farmers and truckers will top $1 billion since Peaster's cow was detected with bovine spongiform encephalopathy on May 20 and the United States slammed shut its border to our beef."

But the U.S. border would prove to have a serious leak. It appears that the United States secretly allowed the importation of millions of tons of Canadian beef into the United States following the Canadian mad cow announcement in May 2003, in spite of public reassurances that imports had stopped. The USDA would admit to the error a year later.

So great was the impact of Canada's mad cow case on the provincial economy that Alberta Premier (Canadian equivalent of state governor) Ralph Klein took aim with both barrels at the owner of the province's infamous mad cow, saying a "self-respecting" rancher would not have taken the animal to slaughter but would instead have simply "shot, shoveled and shut up." Premier Klein would later claim he meant these remarks, for which he was publicly vilified, to be sarcastic. However, the public consensus was that they were remarkably honest and summed up the prevailing, but privately held, attitude in the ranching community.

I spoke with a longtime rancher in Saint Paul, Alberta,

who told me that there is no way a rancher is going to report an animal that died under these circumstances after seeing the misfortune that Marwyn Peaster and his family had to endure. The one and only case of mad cow disease in Alberta will remain that way, regardless of the spread of the disease in the country. There may be thousands of cases of BSE in Canada, but chances are we will never get to hear of them, and neither will the Canadian government. The economic consequences of reporting it are just too devastating, both for the individual and for the economy. It just is not going to happen.

The announcement on May 20 set off one of the most intensive detective hunts in history. The animal was quickly tracked to a location in northwestern Saskatchewan. On May 21, just a day later, the farm of Mel McCrea at Baldwinton, Saskatchewan, and another in the Lloydminster area were placed under quarantine after the Canadian Food Inspection Agency began tracing procedures to locate animals that were fed with, or were offspring of, the BSE cow, as well as farms that may have received potentially BSE-tainted feed. Mr. McCrea told Canada's *Globe and Mail* that he fed his animals 95 percent of his own grown feed plus a protein supplement, which he understood to be rendered-meat free. "My dad's been developing this herd for over forty years," McCrea told the newspaper. "I've worked on them all my life—since I was 10 years old I've been out there helping with the cows. It will be hard to sit here at this kitchen table, look out and not see cows."

Questions were raised as to why it took Canada three and a half months to determine that the eight-year-old Black Angus—slaughtered on January 31 after being pronounced too sickly for human consumption—was infected with mad cow disease, the common name for the mysterious BSE ailment that is caused by an abnormal protein that destroys nervous system tissue.

A timeline constructed in retrospect shows a somewhat casual approach by the Canadian Food Inspection Agency to BSE testing. However, it is hard to lay blame on a government agency that never had reason to suspect a BSE case before. In an attempt to retrace the origins of the BSE case, investigators showed that in 1997 a Black Angus cow was born on one of two farms and in the spring of 1998 the animal was sold to Heartland Livestock Services of Lethbridge, Alberta. According to the timeline, that summer Bryan Babey of Lloydminster, Saskatchewan, bought the cow, and in the four years on the Babey farm, the animal gave birth to four calves. In 2002, Babey sold the cow and a Lloydminister cattle broker sold the animal to the Nilsson brothers in Vermilion, Alberta. On August 23, 2002, Marwyn Peaster bought the animal, and in January 2003 Peaster noticed the animal could not get up, so he shipped it for slaughter. On January 31 the cow was condemned as unfit for human consumption and the head was sent to a lab in Fairview, Alberta. On February 8, part of the cow's head arrived in Edmonton for testing. It was put at a low priority and was placed on a waiting list. On May 12 or 13, tests were done on the brain, and on May 16 the tissue showed a positive BSE test. The sample was then sent to a Canadian Food Inspection Agency lab in Winnipeg, where the BSE result was confirmed, and then it was sent to the U.K. lab for further testing. Meanwhile the Peaster farm was quarantined. On May 20, the

British lab confirmed the BSE in the cow and the CFIA issued its now famous press release.

The ranch where the BSE-positive cow spent considerable time before moving to Alberta was at Baldwinton, a town in northwestern Saskatchewan that lies in the heart of a CWD-endemic area. In recent years, Saskatchewan and, to a lesser extent, Alberta, have been at the epicenter of a prion disease epidemic in deer and elk. Since the first documented case of CWD in Saskatchewan in 1996, about forty separate outbreaks of CWD in farmed deer and elk have been documented, primarily in a region west and northwest of Saskatoon and centered in North Battleford, Saskatchewan. A subsequent investigation showed that authorities were blissfully unaware of the dangers of CWD spreading in Saskatchewan. A Canadian Food Inspection Agency report in July 2003 notes in passing that the BSE case came from this same CWD "hot zone."

But the report contained even more damaging disclosures:

In 1996, a Saskatchewan elk herd became similarly sentinel to the North American presence of CWD in commercially raised cervids. Although the disease had been recognized in wildlife research facilities in Colorado and Wyoming decades before, its ingress into commercial North American elk herds had remained unrecognized until the Canadian investigation signaled its presence. The disease was eliminated from multiple herds in its primary epidemiologic focus in Saskatchewan, concentrated, coincidentally, in the area considered to date as one of the most probable origins

of the index case. The few sites considered heavily contaminated remain cleared of all livestock and wild
cervids. An epidemiologic study of the outbreak's pro-
gression continues. To date, depending on owners' elec-
tion to bury or recycle, the study has identified the possibility that as many as 110 CWD-infected elk
carcasses might have entered the feed distribution clusters shared by the prospective herds of origin of the
index case during the interval 1993–1999 before the
regional renderers ceased acceptance of all TSE-suscep-
tible materials.

Other than the passing reference in the Canadian Food
Inspection Agency report, no one has noted this apparent
coincidence. BSE and CWD appearing in the same area at the
same time? Remarkably, the report says that as many as 110
CWD-infected elk were recycled through the rendering system in the area. Are they saying that the lone BSE "index"
case may have arisen because CWD-infected meat was fed to
the Baldwinton cow? If that is the case, it is almost inevitable
that other cattle were also contaminated.

Another nine cases of CWD in wild deer have been documented in 2002 and 2003 south of that North Battleford
area, primarily in Saskatchewan Landing Provincial Park and
the Sandhills area, although one case of CWD in a wild deer
was reported in Paradise Hill, which lies in the heart of the
CWD epidemic in farmed cervids.

As if to add insult to injury, Canadian CBC News reported
in June 2003 that sheep from Lashburn, Saskatchewan, had
tested positive for scrapie and that the infected sheep had been

sold in Lloydminster. Lloydminster, of course, is the same town that sold the BSE-infected cow. The sheep's owner, Fred Davis, says the sheep came from a flock in Maidstone, Saskatchewan, and that he had bought them in Lloydminster. So now, overlaid on both the CWD epidemic and the BSE outbreak, we have a scrapie outbreak in the same area of northwestern Saskatchewan. The fact that three separate outbreaks of different prion diseases originated in the same area of Saskatchewan between 1996 and 2003 appears to have gone unnoticed. An outbreak of BSE, an outbreak of CWD, and an outbreak of scrapie—all from the same area. Is that coincidence?

Is it coincidence that both Saskatchewan and Alberta were also subject to a wave of cattle mutilations beginning in 1979? The Canadian mutilations were characterized by much less media hysteria than those in the United States and by a much more professional atmosphere. The Royal Canadian Mountain Police were involved from the beginning, as was an extremely capable veterinarian named Dr. David Green. By 1980 Green had conducted almost fifty necropsies on muti-lated cattle and accumulated very strong evidence that the mutilators were using sharp instruments and sedatives, including succinylcholine, to bring down the cattle. Much of Dr. Green's work was never publicized in the United States.

Is it a coincidence that, even before the publicity of either the BSE or the scrapie outbreaks in Saskatchewan, a man from Lloydminster, Saskatchewan, died from sporadic CJD? Andrew Swift of Health Canada said that this CJD death was one of twenty-five to thirty cases that occur in Canada annu-ally. He told CBC News in Canada that CJD is "a naturally occurring disease, with the cases spread randomly across the country." Some months later we would learn that BSE and scrapie had trafficked through Lloydminster as well.

But the first case of variant CJD, or mad cow-related CJD, in Canada also occurred in Saskatchewan. "Fears that the human form of mad-cow disease would strike North America have been realized with the death of a Saskatchewan man from the incurable brain-wasting illness," reported the *Globe and Mail* on August 9, 2003. "The continent's first confirmed case of variant Creutzfeldt-Jakob disease alarmed dozens of people who may have been infected through a medical instrument used on the man, while unsettling untold numbers of Canadians who lived recently in Britain, where the ailment originated."

Government officials in Saskatchewan rushed to reassure the public by claiming that the man had contracted the disease while he was in the United Kingdom. "There is no evidence that mad-cow disease has entered the Canadian food supply and therefore we can reassure the Canadian public that the person did not contract the disease in Canada," said Antonio Giulivi of Health Canada at a news conference, according to an article in the *Globe and Mail*. The people in Saskatchewan must have breathed a sigh of relief, in spite of the fact that Giulivi did not provide any evidence that the Saskatchewan native had contracted variant CJD in the United Kingdom. And just where the variant CJD victim had lived in Saskatchewan was kept secret.

This region of northwestern Saskatchewan, the epicenter of this confluence of CWD, BSE, and scrapie events, is also the epicenter for about eighteen separate cases of animal mutilations investigated since 1994. This is a remarkable overlap between a possible monitoring operation for prion-infected animals and three separate outbreaks of prion disease, all from the same area of Saskatchewan. Similarly, cattle mutilations have been reported in the area of southern Saskatchewan that corresponds to locations where recent CWD cases have been found in wild deer. Therefore, there is a striking geographical overlap between the locations of the CWD, BSE, and scrapie epicenter in Saskatchewan and the locations where animal mutilations were reported since 1994.

I suggest that the animal mutilations reported in northwestern Saskatchewan in the past several years are a covert prion-sampling operation by perpetrators who knew that the infectious agent was spreading from farmed elk and deer in Saskatchewan to wild deer and then to cattle and even sheep. Either we are dealing with an extraordinary coincidence in having three separate outbreaks of prion disease from the same area or the much-vaunted "species barrier" has somehow broken down in this area of Saskatchewan.

It is tempting to speculate that northwestern Saskatchewan became a hot zone sometime in the 1980s and that somebody knew about it. The extraordinary and precise coincidence that three separate prion disease outbreaks occurred in the relatively small area around Lloydminster, Saskatchewan, surely begs the question: Are they connected? And was the overlay of the intensive series of cattle mutilations in the same area of Saskatchewan an "unofficial" monitoring operation of the extent of prion contamination in Saskatchewan cattle? Or was it coincidence?

"Coincidence," said novelist Emma Bull, "is the word we use when we can't see the levers and pulleys."

21

U.S. Mad Cow

Then it was our turn across the border. The first official case of mad cow disease to strike the United States was announced two days before Christmas 2003. The USDA acted quickly on the news of the mad cow in Washington State, but it was obviously too little too late. Within a week the Department of Agriculture's Ann Veneman had announced a ban on the sale of downer cows to the nation's rendering industry. And very quickly, the USDA also announced that the infected cow had come from Canada. The Canadians, already reeling economically from the May 2003 announcement of their own case of mad cow, were not amused. But the detective hunt did appear to point toward Canada as the source of the Mabton mad cow.

In another curious coincidence, it turns out that in the early 1990s, a very intensive program of inoculation of cattle with the brains of mad mink had been ongoing just a few dozen miles from Mabton and Moses Lake, Washington, where the mad cow had been found and slaughtered in December 2003. Pullman, Washington, is the location of one of the USDA's most professional prion testing facilities in the United States. At Pullman, since the early 1990s, a top-gun, award-winning research team led by Dr. Donald Knowles had been inoculating cattle at the Animal Disease Research Unit with mad mink brain slurry obtained from the three most famous mad-mink outbreaks—Stetsonville, Wisconsin; Hayward, Wisconsin; and Blackfoot, Idaho.

The coincidence is a chilling one. Why? Because the outbreaks of mink disease at Stetsonville, Hayward, and Blackfoot had been strongly suspected by Dr. Richard Marsh as originating from a hidden epidemic of BSE in downer cows. Marsh had found very strong, nearly irrefutable evidence that the mink died because they had been fed ground-up remains of BSE-infected downer cows. And here, during 1992–1994, a large-scale inoculation program of injecting cattle with the ground-up brains of the very same mad mink that had potentially died from eating downer cows was being carried out only a few dozen miles from the site in Washington where the first case of "official" BSE would be announced in 2003. The geographical overlap was both stunning and heavily laden with irony.

The ban on the sale of downer cows, which had been blocked in Congress just a few weeks earlier but was now abruptly announced by the USDA, represented, according to the *Washington Post,* "a repudiation of years of industry efforts to limit government intervention in slaughterhouse operations and in shaping the nation's response to the threat of mad cow disease."

In addition to an urgent recall of beef throughout several Western states within days of the announcement of the mad cow case, the USDA also announced restrictions on what could be included in "ground beef." Seven years earlier, the Center for Science in the Public Interest had begun calling for stronger controls to prevent brain and spinal cord tissue from contaminating meat that had been separated from beef bones through a decade-old technology called "advanced meat recovery," or AMR. A survey by the USDA in 2003 found that about 35 percent of ground beef randomly tested in the United States contained "unacceptable levels of brain tissue." Now Veneman was announcing further restrictions on the tissues that could be included in AMR products.

The chickens were coming home to roost after nearly twenty years of denial, beginning in 1985, when Dr. Richard Marsh first began to warn about undetected BSE disease in the nation's cattle herd. That scientists began to rally was an unusual sight, as scientists are not known for coming out with controversial statements, especially those that would conflict with a very powerful lobby. And there are few groups more powerful in the United States than the cattle lobby.

George Carlson, the director of the prestigious McLaughlin Institute in Great Falls, Montana, told the *Great Falls Tribune* that the new government actions to guard against mad cow disease have been recommended for years by scientists and should have been taken before now. He also said government officials weren't being honest when they said there's "zero risk" of contracting the disease by eating regular cuts of meat from an infected animal. These cuts—often called "muscle cuts," such as roasts, chops, and steaks—may contain some of the disease-causing prions more prevalent in brain and spinal tissue.

Even the much-maligned Richard Marsh, who died in 1997, still ignored and ridiculed by the cattle lobby, got some measure of public rehabilitation following the discovery of mad cows in Alberta, Canada, and Washington State. Canada's *Globe and Mail* remarked that many BSE experts now regarded Dr. Marsh's work as prescient. "It was good work," the newspaper quoted Dr. David Westaway as saying. Westaway is a molecular biologist and prion specialist at the Centre for Research in Neurodegenerative Diseases at the University of Toronto. "It was ignored unfairly, and it was years ahead of its time," he said. Westaway stated that the current system was anything but science based, and noted that the U.S. Cattlemen's Association had been "virulently" against testing, a position that had influenced Canadian policy.

A month after the announcement of the Department of Agriculture's restrictions on selling downer cows to the rendering industry, there was a much ballyhooed announcement from the FDA, saying that the agency would move to close three of the most egregious loopholes in the alleged 1997 "firewall" imposed by the USDA on feeding cattle parts to other cattle. The three loopholes were

1. the practice of feeding poultry litter (excrement) to cattle because poultry feed itself includes cattle remains
2. the feeding of restaurant waste (which contains beef) to cattle and
3. feeding cattle and sheep blood to cattle and calves.

The latter practice is common, and particularly in the veal industry, cattle and sheep blood are force-fed to young calves to bulk them, using the euphemistic term "milk replacer."

Most consumer groups applauded the FDA's announcement that these three loopholes would be plugged. But three months later, the rules still had not reached the Federal Register, a required step before a rule is acted upon. And since they had not reached the mandatory stage, farmers and the industry were not required to follow them. So in the spring of 2004 it was entirely possible that cattle throughout the United States were still being fed cattle parts and the public was still eating potentially tainted beef. In fact, the FDA did not move toward a ban on feeding farm animal parts to farm animals until the summer of 2004. Consumer groups criticized the move as it left open the loophole of feeding cattle the blood of processed cattle and other materials that might spread the disease.

Meanwhile, another troubling revelation had hit the news. In February, questions had been raised about the trustworthiness of the mad cow testing process at the central testing facility in Ames, Iowa, known as NVSL. "Distrust of the NVSL is so widespread among USDA veterinarians and meat inspectors it limits mad cow disease surveillance 'tremendously,' " a veterinarian with more than twenty-five years of experience with the agency told Steve Mitchell of UPI. The veterinarian had requested anonymity because he feared repercussions. But he went on to tell Mitchell that many agency inspectors don't bother to inspect cows closely for signs of mad cow disease or to send brain samples to the lab because there is little chance it will issue a positive result—even if the cow is infected.

The alarming part of Mitchell's story was that even other

veterinarians in the business have stated that undoubtedly other cases of mad cow disease had already been detected in the United States, but had gone undisclosed through incompetence or otherwise. The veterinarian went on to tell Mitchell that "Most agency veterinarians know mad cow is prevalent and epidemic (in U.S. herds). We're not talking about one or two cases."

In March, a local newspaper in Ames, Iowa, reported that the testing facility was being moved from its present location in a strip mall in Ames and that samples of potentially BSE-contaminated meat were being removed from the premises because of lax security. A report by the inspector general's office stated that not only was the facility close to other businesses, but it "has limited security at its entrances and exits."

The question of whether Washington's mad cow was a downer, or was still healthy when tested, still raged three months after the discovery. At stake was a simple yet profound issue. The USDA position was that the cow was killed because she was a downer and then, following USDA protocols, inspection of the brain showed that the animal had mad cow disease (BSE). An entirely different and much more chilling story emerged from Dave Louthan, the person who actually put a bullet through the poor animal's head. Louthan insisted the cow was perfectly healthy-looking.

Then, in the first week of March 2004, the inspector general's office of USDA announced that a criminal probe was being conducted into whether the records were falsified in order to state the cow was a downer, not a walker. Phyllis Fong, the inspector general, told a House appropriations subcommittee that in addition to the criminal investigation, a second investigation into the USDA's response to the mad cow case was underway. "There's a culture in the industry that is fostered by an inappropriate relationship between the government and the cattle industry," said Susan Solarz, a biology scholar-in-residence at American University in Washington, D.C., in a story in *The Sacramento Bee*, an award-winning newspaper known for breaking big stories. "There are so many people with ties to cattle in decision-making positions in the USDA that there is basically an unofficial agreement to not test more animals. If this was not a downer, then it means that the USDA's policy was seriously flawed."

Rodney Thompson is the veterinarian who conducted the testing on Washington's mad cow on December 9 outside Vern's meatpacking facility. Thompson is the only USDA witness who says the cow was unable to stand or walk. Three other witnesses who saw the cow the day it was slaughtered have said that the cow was walking. Thompson has made no public statement about whether the cow was a downer. The USDA has kept the press away from Thompson and refuses to provide any information about him.

Louthan told UPI's Steve Mitchell that Thompson was "a good and honorable man" and that the easiest explanation was that the USDA had ordered Thompson to falsify the evidence—or else they had falsified the evidence after Thompson had submitted his written records. Before the announcement of the criminal probe, Louthan began complaining that USDA officials were threatening him to keep him quiet. Louthan said that he had been forced into a car outside his home by two agency officials and told to keep quiet. The USDA denies he was intimidated.

Despite the alarming situation, the domestic beef and dairy market appeared to be holding strong. The campaign by the USDA and the cattle lobby had been extremely effective. Just to underscore the general apathy in the American public, CNN in mid-January ran a story called "Cow Brain Sandwiches Still on Some Menus." "Fear of mad cow disease hasn't kept Cecelia Coan from eating her beloved deep-fried cow brain sandwiches," read the CNN report. "She's more concerned about what the cholesterol will do to her heart than suffering the brain-wasting disease found in a cow in Washington State. 'I think I'll have hardening of the arteries before I have mad cow disease,' said Cecelia Coan, 40, picking up a brain sandwich to go at the Hilltop Inn during her lunch hour. 'This is better than snail, better than sushi, better than a lot of different delicacies.'"

The twin actors in this drama, the USDA and the cattle industry, working in unison like a well-oiled machine, had persuaded the American public that eating beef was safe—safe enough even to eat cow brain sandwiches! In spite of the announcement of a criminal probe into the falsification of the records of the mad cow case, the American public continued eating beef. And in a further move, only days after the criminal probe began, the USDA announced that it would be testing 120,000 of the nation's 35 million animals slaughtered. This headline was the coup de grâce. The domestic market, in spite of a couple of speed bumps, had remained stable.

The export market, however, was a different story. Fifty-eight countries had shut their borders to American beef. As the months following the American mad cow case continued and still the export ban to other countries, particularly Japan, showed no signs of abating, many of the smaller slaughterhouses in the United States became increasingly alarmed. Creekstone Farms Premium Beef of Kentucky wanted to use recently approved rapid tests so it could resume selling its fat-marbled Black Angus beef to Japan. Creekstone calculated that the Japanese ban was costing the company about $40,000 per week, and the plant was going to have to lay off about fifty workers. Creekstone hit on an idea. Since Japan had always tested 100 percent of the cows that the public ate, they reasoned, why not emulate the Japanese system for every cow that went through the rendering plant? (Europeans now test about half of all cattle that are consumed by the public.) Why not test the cattle for mad cow disease and inform Japan of this new policy; that way all problems would be solved. Japan agreed to the policy.

The USDA did not. The USDA refused to let the plan go ahead, saying that for Creekstone to test every cow would be "unscientific." Bill Hawks, the department's undersecretary for marketing and regulation, explained its decision, saying that the rapid tests were only licensed for surveillance of animal health and that Creekstone's use of the tests would have "implied a consumer safety aspect that is not scientifically warranted." The problem dates back to 1913, when Congress passed the Virus Serum Toxin Act. The act gave the USDA the power to tell ranchers and meatpackers what safety tests they had to perform. Remarkably, it also gave the USDA the ability to *prevent* testing. "In an abuse of this 91-year-old law," said an editorial in *The Sacramento Bee*, "the USDA has told a Kansas beef producer that it can't test every slaughtered animal for mad cow disease."

A month later, a second slaughterhouse, this time in northwestern Alberta, announced that it, too, would begin testing all slaughtered beef and it would institute a policy of bar coding to be able to track the origin of every animal. This system has been in place in Europe for years, but again due to strong resistance from the cattle industry, it has not been instituted in North America. The Canadian economy in particular had suffered deeply since the United States and other countries had closed their borders to Canadian beef. In Canada, the move in Alberta was seen as a way to escape the economic restrictions on the ban on Canadian beef. Since the ban had been imposed on Canadian beef, an estimated $2 billion dollars had been lost to the Canadian economy.

In early May 2004, a potential second case of mad cow disease appeared in Texas, but the USDA quickly snuffed it out. "The cow, which staggered and collapsed after passing an initial visual inspection at Lone Star Beef in San Angelo, Tex.," reported *The New York Times*, "was condemned as unfit for human consumption and under federal regulations should have been tested for mad cow disease. Instead, it was sent to a rendering plant to be made into animal food and byproducts." The incident led to several calls for a federal inquiry as to why the USDA did not test a cow that had collapsed with neurological symptoms. *The New York Times* then noted that "Consumer groups have regularly accused the Agriculture Department of trying to avoid finding more mad cow disease because of the damage it would do to the beef industry. Former beef industry officials hold high positions in the department."

An investigation by Steve Mitchell of UPI revealed that the USDA tested only three animals out of a total of 350,000 cattle that had been slaughtered at the Lone Star Beef facility in 2002 and 2003. It later transpired that the Lone Star facility tended to slaughter animals that were older than thirty months; older animals have a much higher probability of having mad cow disease. Lone Star, however, said that the low testing numbers were explained by the fact that the plant does not accept downer cows, and that the USDA normally tests only downers. This may be a good indication of why the USDA had only publicly announced a single case of mad cow disease in the United States by the spring of 2004.

On June 1, 2004, the USDA initiated an intensive mad cow surveillance policy by subjecting up to 220,000 downer cows to a new rapid screening test that only took four hours. The aim was to rapidly test up to 220,000 downer cows in the United States over an eighteen-month period. A month later, the USDA had announced that two animals tested were positive for mad cow in the preliminary rapid test, but subsequent testing using the IHC "gold standard" assay had found both animals negative. Reporting on the latest testing, UPI's Steve Mitchell found a food industry consultant who estimated that "there could be more than 100 cases of the deadly disorder [mad cow disease] in the country's herds."

Across the Atlantic Ocean, a newly released French government study showed that a mad cow disease epidemic in the 1980s had gone completely undetected in France and that an estimated 50,000 severely infected animals had entered the French food chain. Simultaneous with these disturbing findings, officials announced that a seventh person in France had died of variant Creutzfeldt-Jakob disease.

The truth was now slowly emerging. The question at this time is: Is the mad cow situation about to take a turn for the worse on both sides of the Atlantic?

22 Origins

This, then, is how I think the prion catastrophe occurred in the United States. It likely began with the routine importation of dozens of kuru brains from New Guinea by Carleton Gajdusek and Joseph Smadel beginning in 1957. Subsequently, there were large-scale inoculations of that kuru material into dozens of species of animals in the middle of a wildlife refuge at Patuxent, Maryland, from 1963 until 1970. "They even inoculated alligators" is a phrase that sums up their indiscriminate approach.

The importation of dozens of kuru-loaded brains from the wilds of New Guinea into Maryland may have been the first step in the spread of infectious prions into the wildlife population in the United States. The second step was the frantic, widespread, but inadvertent amplification of prion diseases through multiple species. It is a fact that the containment facilities at Patuxent, which were shared with groups from

Walter Reed Hospital and perhaps Johns Hopkins University among others, were poor enough that at least one attempt was made to shut them down. That attempt by Robert C. Reisinger of the Animal Disease Eradication Division at the USDA was ultimately unsuccessful. The Patuxent facilities originally consisted of a few disused buildings in the middle of a well-stocked wildlife refuge where deer grazed, birds migrated, and dozens of species of wildlife passed through. There were open runs where inoculated primates moved and defecated. My attempts to discuss the possibility of escapes from Patuxent were met with a short email reply from Dr. Paul Brown, a recently retired neuropathologist at NIH and a colleague of Gajdusek's: "No animals escaped from Patuxent. Period." Dr. Brown's statement does not agree with Dr. Gibbs's admission to Rhodes that the chimpanzees at Patuxent opened the cage doors and allowed the monkeys to escape (they were recaptured a short time later). But in addition to the inoculations carried out at Patuxent, other inoculations took place at facilities in Louisiana and Washington State.

Did the prion diseases that are threatening our wildlife originate with these inoculations? Were they fuel on the fire of prion diseases that were already present in the wild? There is no conclusive proof. What is clear, however, is that in the years that followed, prion disease spread slowly and insidiously throughout North America. The disease appeared in mule deer in the wildlife research facilities at Fort Collins, Colorado, and its sister institute in Wyoming in 1967, possibly as a result of scrapie-infected sheep being housed in the same facilities. But the disease may also have erupted from contaminated wildlife in the area around the facilities. Whether this epidemic in the mule deer population was as a result of, or independent of, the kuru experiments will probably never be determined. In any case, it is difficult to refute the fact that the Colorado and Wyoming wildlife facilities subsequently became the epicenter for an epidemic of chronic wasting disease, which rapidly spread into neighboring states, beginning with Nebraska and moving into the rest of Wyoming.

Chronic wasting disease then jumped from northern Colorado to within a few miles of the Mexican border in White Sands, New Mexico. Wildlife officials have ruled out transport of infected deer from elsewhere and they have ruled out deer migration patterns. The bizarre jump of CWD into southern New Mexico and the middle of Utah led CWD experts to state, in a journal called *Emerging Infectious Diseases,* that *". . . unidentified risk factors* [my italics] may be contributing to the occurrence of CWD among free-ranging and captive cervid populations in some areas." What could these be? A massive, pervasive, and presently unseen epidemic of prion disease in wildlife is one obvious explanation for these bizarre "jumps."

Again, deer are probably only the sentinel species in this epidemic. Nobody is spending a lot of time looking for raccoons or squirrels with unsteady walks. But that doesn't mean the disease isn't striking those species as well. And, in fact, it has. Prion disease erupted in squirrels in Kentucky in the 1990s. Although the data are dismissed as "soft," five people died of CJD after eating squirrel brains in Kentucky during this time. And in 1985 there was a massive outbreak of prion disease in mink in Wisconsin. That epidemic is even more worrying because, as Dr. Richard Marsh pointed out, the prions may have first transitioned into Wisconsin mink from cattle. While that disturbing scenario is still being debated, it is undeniable that the outbreaks in Kentucky and in Wisconsin signaled a spread.

The idea of a "species barrier" that stops the spread of prions between different species is comforting, but it's not scientific. The British Ministry of Health and Agriculture successfully comforted the British populace for almost a decade with the skillful use of the phrase "species barrier," saying in effect that humans could never become ill from eating mad cows because of the species barrier. Unfortunately, reality intruded into this idyllic picture and scores of young people began dying from mad cow disease. In 2004, the existence of a prion species barrier is more an expression of politics than of science.

Scientific research has now shown that the notion of a "species barrier" firewall was a hope and a dream, a smoke-and-mirror job to reassure the public and prevent panic from spreading. Two major studies published in April 2004 provide some insight into the mechanism by which prions of one species can invade another. All it takes, apparently, are prions from a third species to form a template or stencil that will hasten a shape change and a subsequent jump from one species to another. And each time the jump is made from one species to the next, the greater the prion's potential to invade still other species, in some cases with greater virulence.

This research has major implications for the further spread of disease that policy makers shouldn't ignore. "The bottom line is, if we don't tightly control these diseases, we're going to regret it big time," Dr. Pierluigi Gambetti, director of the

National Prion Disease Pathology Surveillance Center at Case Western Reserve, told the *Cleveland Plain Dealer*. They "may come in through the back door."

The prion's ability to adapt and exploit a weakness in the species barrier has surprised everyone. "It is reasonable to feel a little less safe," said Dr. Neil Cashman, a neurologist at the University of Toronto's Center for Research in Neurodegenerative Diseases. "This study points up the fact that our previous reassurances are wrong. But it also provides a signal that much more science is needed. There are many things we don't understand, and the whole science of how prions propagate and cross species barriers is developing as we speak."

This latest research throws much light on how the disease may have gotten out into the wildlife population of the United States after the inoculations of literally dozens of species at the Patuxent wildlife facility in Maryland in the early 1960s. Second, it provides a rationale for the emergence of CWD into the Colorado area from sheep or other wildlife. And third, it strongly contradicts experts who claim how highly unlikely it is for the CWD epidemic in a dozen states to pass to cattle and humans.

Just how easy it is for one prion disease to spread among wildlife cannot be underestimated and is only now being realized. At the University of Wisconsin, Dr. Michael Samuel has been staking out deer carcasses to study exactly which scavengers come to the decomposing body to eat. Using flash-lit photography, Dr. Samuel and colleagues have already seen a much wider variety of species feeding on the deer carcass than anticipated, including hawks, owls, crows, dogs, cats, coyotes, raccoons, skunks, mink, foxes, and opossums.

Think of this happening at Patuxent forty years ago: A kuru-loaded animal (from whatever species) escapes and dies and the carcass is scavenged on by another dozen species. Can some of these wild species actually be infected with prion diseases? Studies have already shown that prion disease in both mink and deer can be transmitted experimentally to raccoons. And birds can get prion disease as well; an outbreak occurred in a flock of ostriches at a German zoo. Numerous studies have also shown that cats have died as a result of eating prion-contaminated food. And most disturbingly, recent studies have shown that flies could act as vectors and even reservoirs for the spread of prion disease. Could prion disease have escaped from Patuxent and spread to local wildlife via flies?

I have not been able to find any records of such an occurrence at Patuxent, despite the fact that Dr. Gibbs admitted to at least one escape. If, as was alleged by Reisinger, the containment at Patuxent was inadequate, the inoculations of thousands of animals at Patuxent became, albeit unwittingly, the equivalent of a prion-loaded nuclear reactor in the pristine Maryland countryside. If there was an escape by an infected animal, wildlife in the area could have become contaminated; the contamination could have spread.

The cattle mutilation phenomenon emerged in the United States after the massive kuru inoculation program at Patuxent had been underway for several years. From all appearances, the phenomenon represents a largely covert monitoring operation of the spread of prion disease into our nation's food supply. Was somebody in those early days concerned about the consequences of kuru spreading through the human food chain? Was someone being extraordinarily prescient? After all, back then nothing was known about prions. But a good deal of information was, in fact, known about the deadly "slow virus" diseases.

It was known, for instance, that kuru had wiped out a few generations of Fore tribespeople in the 1940s and the 1950s. The devastation of the Fore tribe via kuru was a dramatic example that the infectious agent was transmissible. The kuru catastrophe in New Guinea as documented by Gajdusek even made *Time* magazine back in the late 1950s. It was big news.

That cattle might be at risk was also being discussed at the National Institutes of Health. Dr. Gaylord Hartsough had raised the awareness about a possible cattle infection in December 1964. He presented data at a very influential gathering of the top scrapie and kuru researchers in the world and his contention would have been taken seriously, if not by Gajdusek and his cohorts, then certainly by other attendees. For example, Robert C. Reisinger, the veterinarian who made strenuous efforts to shut down the Patuxent facility the previous year, also attended that NIH conference. There was undeniable fear on Reisinger's part, and by others as well, that this disease, about which so little was known, could end up infecting the agriculture sector of the United States. It was already known to infect sheep in the guise of scrapie, and here was Hartsough presenting data that maybe cows were affected as well.

In any case, by the late 1960s the mysterious killing and sampling of cattle had quietly begun. Ranchers were outraged. Their animals were being ruthlessly killed and samples of their eyes, tongues, reproductive organs, and intestines were being silently carted off under cover of darkness. Sedatives or tranquilizers were used to subdue the animals. Often sharp instruments were used to remove tissues from these freshly killed cattle. And the very organs that were removed—the eye, the tongue, the reproductive organs, and the large intestine—are certainly organs that, in hindsight, have been shown to be early prion reservoirs.

The monitoring operation was definitely illegal. It was ruthless in its efficiency. It had no regard for the wishes of ranchers. Top-gun operatives swooped in at the dead of night and sampled animals from a wide variety of hot zones. No footprints or tracks of vehicles were left behind. Helicopter activity has always been associated with these operations. It was, in other words, highly efficient.

These attributes do not identify those responsible for the monitoring operation. The "monitor group" would not have to be especially large to have carried out painstaking research all over the United States to monitor the spread of prions through the human food chain. It could simply be a private organization on contract, a sanctioned or unsanctioned government department, or another private organization with funding. In 2004, much has been made of the "cowboy," and sometimes illegal, operations carried out by private contractors in Iraq. Private contractors have been proliferating exponentially since the 1950s in the United States, so the concept of illegally killing, mutilating, and obtaining organs from cattle is certainly not beyond possibility, given the systemic abuses that have been exposed over the past fifty years. Cattle mutilations also could be used to monitor the spread of a wide variety of infectious diseases through the nation's herds, in addition to prions.

What we do know about the perpetrators is that they are very good at their job. High levels of stealth and skill were used in avoiding capture. In fact, they have never been caught in spite of sampling animals from almost every state in the union at different times over the past thirty-five years.

The two centers of the cattle mutilation phenomenon that are of greatest interest during this time are northeastern Colorado, where the CWD epidemic originated, and northwestern Saskatchewan, where three separate outbreaks of prion diseases (CWD, BSE, and scrapie) have erupted. It is my contention that the extraordinary overlap between cattle mutilations and prion disease in northeastern Colorado and in northwestern Saskatchewan may not be coincidence.

When Governor Richard Lamm flew to Pueblo, Colorado, to meet with cattle officials in 1975, he proclaimed that cattle mutilations were one of the greatest outrages of the twentieth century. In 1974 and 1975 hundreds of animals had died in the area north and northeast of Denver. Within a few short years the same area would become the epicenter of the rapidly expanding CWD outbreak. The cattle mutilations were also reported, though to a lesser degree than in Colorado, in Nebraska and Wyoming, as was the CWD epidemic as it moved north and east. Even in the late 1970s, there was definite concern about the possibility that this disease in deer might affect cattle.

The overlap between the spread of cattle mutilations and the spread of CWD is undeniable. But it goes further than that. The fact is that deer and elk have been mutilated in exactly the same way as cattle. The same organs have been removed. These mutilations are completely different from the standard field dressings that hunters use to remove meat from deer or elk carcasses. With mutilations only the organs are taken, the meat is ignored.

The other cattle-mutilation hot zone occurred in Saskatchewan in the 1990s, during which there were twenty documented cases of cattle surgeries. The mutilations, which centered on Lloydminster, North Battleford, and places north, are the precise locations where years later three separate outbreaks of prion disease were discovered. First, beginning in 1996, CWD broke out in about forty separate areas focused around northwestern Saskatchewan; then, in 2003, an outbreak of BSE occurred in Baldwinton; and finally, in 2003, an outbreak of scrapie happened in the Lloydminster area. Coincidentally, that same area of Saskatchewan is the site of Canada's one and only officially recognized case of variant CJD. If the individual did not contract the disease in Britain, as officials claim, then that makes a fourth outbreak of prion disease in the same area of Saskatchewan, making northwest Saskatchewan a true prion "hot zone."

We are now standing at an extraordinary crossroads. It has been almost fifty years since the importation of infectious kuru into North America. During that period, deadly prions have popped up occasionally in wildlife and now in cattle. Like ghostly wraiths, they have moved silently between species. Only in 2004 has science begun to show that the "firewalls" between species are capable of being breached.

The "species barrier" is actually nothing more than a phantom, a chimera, an illusion. Now, if the "clusters" of CJD are found to be actual clusters, this deadly phantom's attack on humans will be confirmed.

Meanwhile, the massive epidemic of Alzheimer's disease threatens the entire healthcare system in North America. And hidden in that epidemic are the unmistakable signs of a large number of CJD cases that are still unrecognized. Not only are physicians prone to misdiagnose CJD as Alzheimer's disease, but contrary to pronouncements by health officials there is accumulating scientific evidence suggesting that an unknown percentage of so-called sporadic CJD cases may actually be caused by eating contaminated meat.

It took the British government eight long years—from April 1985 with Jonquil, the first cow to come down with BSE, to March 1993, when officials announced a new fatal variant of mad cow disease in humans—to acknowledge the unfolding catastrophe. There are many lessons to be learned from a close scrutiny of the events in Britain. But so far the reactions from the Canadian and United States governments have not been encouraging. The situation in North America is potentially more serious in 2004 than in the United Kingdom in the 1990s, because overlaid on a silent BSE infection in the nation's cattle is an overt and rampantly spreading epidemic of CWD in the nation's deer and elk. The Europeans did not have to contend with a simultaneous wildlife epidemic. And by some estimates the extent of the CWD epidemic in North America is approaching that of the mad cow crisis at its height in Britain in the early 1990s. How can we look at the devastation wrought in Britain and Europe and do nothing?

The bottom line is this: Unless the Centers for Disease Control, the National Institutes of Heath, and the U.S. Department of Agriculture join forces to urgently monitor both these epidemics, BSE and CWD, by the end of the first decade of the twenty-first century we may be faced with a public health emergency of unimaginable proportions.

Given the potential of a looming public health catastrophe, U.S. and Canadian health authorities should undertake immediate action. First, they should dramatically scale up the testing of North American cattle for BSE. Currently, Japan tests 100 percent of all cattle that consumers eat and Europe tests on average about 50 percent of its cattle. The current USDA and cattle lobby position that such a huge increase in testing would be "unscientific" is largely a smokescreen designed to hide their current policy of "don't look, don't tell." The success of the monitoring in Europe and Japan has led to a large number of BSE-positive cattle being detected. As of May 14, 2004, the cumulative BSE-positive cases worldwide are as follows: United Kingdom 182,547; Austria 1; Belgium 124; Czech Republic 11; Denmark 13; Finland 1; France 919; Germany 319; Greece 1; Ireland 1,417; Israel 1; Italy 121; Japan 11; Liechtenstein 2; Luxembourg 2; Netherlands 76; Portugal 879; Poland 14; Slovakia 15; Slovenia 4; Spain 428; Switzerland 453; Canada 1; United States 1. Given the accumulating circumstantial evidence that BSE may be present and undetected in both the United States and Canada, the fact that only one BSE case has been officially reported in each country is simply not credible. I believe that public health concerns would be assuaged if a truly random monitoring program were instituted in North America that tested just 20 percent of all cattle destined for consumers. This testing program should incorporate random audits and quality control testing independent of the current USDA system because of the real or perceived conflict of interest between the cattle industry and the USDA.

Second, the United States and Canada need to institute nationwide mandatory testing programs for CWD, not only in states and provinces that have reported CWD, but in those that have not. Individual wildlife departments in each state currently afflicted with CWD have made great strides in testing deer and elk in 2003 and 2004. In addition, an aggressive epidemiological survey should be instituted immediately of all deer hunters and venison eaters to track those suffering from "dementia." This program should begin in Colorado and Wyoming, and continue in Wisconsin, Saskatchewan, and Alberta.

Third, the United States should begin a high-profile, nationwide, mandatory program of CJD surveillance that will cover the full spectrum of "dementia," Alzheimer's disease, and CJD. Adequate funding should be allocated for this program and all of the healthcare agencies, including the NIH, CDC, and the USDA should have cross-disciplinary sharing of databases, in much the same way that the Department of Homeland Security has forced formerly independent agencies (FBI, CIA, DIA) to talk to one another in an unprecedented manner. A special emphasis should be given to investigating apparent or alleged clusters of dementia, Alzheimer's disease, and CJD. Furthermore, the testing and investigation of all "early onset" Alzheimer's disease

(Alzheimer's disease in persons younger than sixty-five) for evidence of prion association should be given a high priority.

Fourth, as was done recently in the United Kingdom, a massive, retrospective random sampling of surgically removed tonsil and appendix tissues should be conducted throughout North America as soon as possible. This will provide us with a good statistical estimate of the degree of infiltration of the deadly prions into the human population.

Fifth, a much higher profile public awareness campaign should be instituted about the passage of deadly CJD through surgical instruments, dental instruments, and, of course, blood transfusions. The public should be made aware of the dangers of the accidental spread of prions through surgery, and particularly through dental procedures, given the volume of new data that prion reservoirs exist in the tongue, the buccal cavity, and the tonsils. The mouth is a potentially prion-loaded site and dental procedures should reflect this situation.

And sixth, an official nationwide testing program should be instituted to test wildlife for the spread of prions and to assess neurological damage. In addition to the current widespread testing programs on deer and elk, road-killed raccoons, skunks, chipmunks, foxes, mink, squirrels, armadillos, frogs, toads, beavers, muskrats, woodchucks, otters, porcupines, coyotes, rabbits, rats, birds, opossums, dogs, and cats should be tested as well. The roadkill-monitoring programs already in effect in some states could be linked to a roadkill prion- and pathology-testing program.

There are a number of factors in our favor that prevent us from being completely pessimistic in the face of this potential public health emergency. We have a very active citizens consumer network that is monitoring on a daily basis every mistake and policy change at both the USDA and the CDC and instantly publicizing it on the Internet. We have a few journalists who are conducting high-quality investigations in the grand tradition of investigative journalism, and their stories, too, are being followed worldwide. In May and June 2004, partly due to the relentless attacks by consumer groups, there appears to be increasing recognition at the USDA that the policy of "don't look, don't tell" is not working. Even Secretary of Agriculture Ann Veneman has admitted she expects to find "a few" more BSE cases soon. And lastly, we have the recent openness by the British government, which placed its files, warts and all, covering almost twenty years of their work on the BSE tragedy, onto the Internet for the world to learn the lessons from their mistakes. If our government acts now to discover the scope and extent of the prion catastrophe in our midst, we may still have cause for hope.

It is not too late.

References

1: The End

Mitchell, Steve, "How USDA Detected First U.S. Mad Cow Case," UPI, January 9, 2004.

Schlosser, Eric, "The Cow Jumped Over the USDA," *The New York Times*, January 2, 2004.

Ostrom, Carol M., "Worker Says Discovery of Infected Cow Was 'a Fluke,'" *Seattle Times*, January 24, 2004.

Sleeth, Peter, and Andy Dworkin, "Cow's 'Downer' Status Comes into Question," *The Oregonian*, January 23, 2004.

2: Kuru

Zigas, Vincent, *Laughing Death: The Untold Story of Kuru,* Clifton, NJ: Humana, 1990.

Gajdusek, D. Carleton, "Autobiography," Nobel Foundation, 1976, http://www.nobel.se/medicine/laureates/1976/gajdusek-autobio.html.

Hooper, Edward, "Dephlogistication, Imperial Display, Apes,

Angels and the Return of Monsieur Emile Zola," in *Origin of HIV and Emerging Persistent Viruses* (Proceedings of a Conference, *Atti dei Convegni Lincei,* Rome, September 28–29, 2001): Vol. 187 (2003). http://www.uow.edu.au/arts/sts/bmartin/dissent/documents/AIDS/Hooper03.pdf.

Hooper, Edward, *The River: A Journey to the Source of HIV and AIDS,* Boston: Little, Brown, 1999.

Gajdusek, D. Carleton, "Unconventional Viruses and the Origin and Disappearance of Kuru," Nobel Lecture, December 13, 1976. http://www.nobel.se/medicine/laureates/1976/gajdusek-lecture.pdf.

Gajdusek, D. Carleton, *Correspondence on the Discovery and Original Investigations on Kuru: Smadel-Gajdusek Correspondence, 1955–1958,* DHEW Publication No. (NIH) 76-1168. U.S. Department of Health, Education and Welfare, 2nd printing, 1976.

Rhodes, Richard, *Deadly Feasts: Tracking the Secrets of a Terrifying New Plague*, New York: Touchstone, 1998.

3: First Link

Klatzo, Igor, D. Carleton Gajdusek, and Vincent Zigas, "Pathology of Kuru," *Laboratory Investigation*, 8, no. 4 (1959): 799–847.

Creutzfeldt, Hans Gerhard, "Über eine eigenartige herdfôrmige Erkrankung des Zentralnervensystems," *Zeitschrift fur die gesamte Neurologie und Psychiatrie* 57 (1920):1–18. Full English translation published as "On a Particular Focal Disease of the Central Nervous System (Preliminary Communication)," in *Alzheimer Disease & Associated Disorders* 3, no. 1/2 (1989): 15–25.

Zigas, Vincent, and D. Carleton Gajdusek, "Kuru: Clinical Study of a New Syndrome Resembling Paralysis Agitans in Natives of the Eastern Highlands of Australian New Guinea," *Medical Journal of Australia*, 44, no. 21: (1957) 745–54.

Gajdusek, D. Carleton, and Vincent Zigas, "Degenerative Disease of the Central Nervous System in New Guinea—The Endemic Occurrence of "Kuru" in the Native Population," *New England Journal of Medicine* 257 no. 20: (1957): 974–78.

Gajdusek, D. Carleton, *Correspondence on the Discovery and Original Investigations on Kuru: Smadel-Gajdusek Correspondence 1955–1958.* DHEW Publication No. (NIH) 76-1168. U.S. Department of Health Education and Welfare, 2nd printing, 1976.

4: Mad Sheep

Brown, P., and R. Bradley, "1755 and All That: A Historical Primer of Transmissible Spongiform Encephalopathy," *British Medical Journal* 317 (1988): 1688–92.

McFadyean, John, "Scrapie," *Journal of Comparative Pathology and Therapeutics* 31 (1918): 102–31.

Schwartz, Maxime, *How the Cows Turned Mad,* Berkeley: University of California Press, 2003.

Cuillé, Jean, and Paul-Louis Chelle, "Pathologie animale. La maladie dite de la tremblante du mouton est-elle inoculable?" *Comptes Rendus de l'Academie des Sciences* 203 (1936): 1552–54.

Cuillé, Jean, and Paul-Louis Chelle, "La tremblant du mouton est bien inoculable/Sheep scrapie is inoculable," *Comptes Rendus de l'Academie des Sciences* 206 (1938): 1687–88.

Pattison, Iain H., W. S. Gordon, and G. C. Millson, "Experimental Production of Scrapie in Goats," *Journal of Comparative Pathology* 69 (1959): 300–13.

Pattison, Iain H., and Geoffrey C. Millson, "Distribution of the Scrapie Agent in the Tissues of Experimentally Inoculated Goats," *Journal Comparative Pathology* 72 (1962): 233–44.

Hadlow, William J., "Scrapie and Kuru," *The Lancet* 2 (1959): 289–90.

5: Breakthrough

Gajdusek, D. Carleton, *Correspondence on the History and Original Investigations on Kuru: Smadel-Gajdusek Correspondence 1955–1958,* DHEW Publication No. (NIH) 76–1168. U.S. Department of Health Education and Welfare, 2nd printing, 1976.

Rhodes, Richard, *Deadly Feasts: Tracking the Secrets of a Terrifying New Plague,* New York: Touchstone, 1998.

Chandler, Richard L., "Encephalopathy in Mice Produced by Inoculation with Scrapie Brain Material," *The Lancet* 1 (1961): 1378–79.

Sigurdsson, Björn, H. Grimsson, and P. A. Palsson, "Maedi, a Chronic, Progressive Infection of Sheep's Lungs," *Journal of Infectious Diseases* 90, no. 3 (1952): 233–41.

Klitzman, Robert, *The Trembling Mountain. A Personal Account of Kuru, Cannibals and Mad Cow Disease,* New York: Plenum, 1998.

Filed Memoranda APHIS Archives: Animal Disease Eradication Division, ARS (1957–65); Animal Health Division, ARS (1965–70, to Veterinary Services); Animal Inspection and Quarantine Division, ARS (1957–65), located at National Archives Administration, College Park, Maryland.

Gibbs, Clarence Joe, D. Carleton Gajdusek, D. M. Asher, et al., "Creutzfeldt-Jakob Disease (Spongiform Encephalopathy): Transmission to the Chimpanzee," *Science* 161 (July 26, 1968): 388–89.

6: Mad Mink

Hadlow, William J., "Reflections on the Transmissible Spongiform Encephalopathies," *Veterinary Pathology,* 36 (1999): 523–29.

Eckroade, R. J., G. M. Zurhein, R. F. Marsh, et al., "Transmissible Mink Encephalopathy: Experimental Transmission to the Squirrel Monkey," *Science* 169 (1970): 1088–90.

Hanson, Robert P., R. J. Eckroade, R. F. Marsh, et al., "Susceptibility of Mink to Sheep Scrapie," *Science* 172 (1971): 859–61.

Marsh, Richard F., and R. H. Kimberlin, "Comparison of Scrapie
and Transmissible Mink Encephalopathy in Hamsters. II. Clinical Signs, Pathology, and Pathogenesis," *Journal of Infectious
Diseases* 131 (1975): 104–10.

Marsh, Richard F., "Transmissible Mink Encephalopathy," in *Prion
Disease of Humans and Animals* (S. B. Prusiner, ed.), Chichester,
England: Ellis Horwood, 1993: 299–306.

Burger, Dieter, and Gaylord Hartsough, "Transmissible Encephalopa-
thy of Mink," Monograph No. 2, Workshop and Symposium on
Slow, Latent, and Temperate Virus Infections, NIH, Maryland,
December 7–9, 1964.

7: Cannibalism

Rhodes, Richard, *Deadly Feasts: Tracking the Secrets of a Terrifying
New Plague*, New York: Touchstone, 1998.

Gajdusek, D. Carleton, *Correspondence on the Discovery and Original
Investigations on Kuru: Smadel-Gajdusek Correspondence,
1955–1958*, DHEW Publication No. (NIH) 76–1168. U.S.
Department of Health Education and Welfare, 2nd printing,
1976.

Lindenbaum, Shirley, "Kuru, Prions and Human Affairs: Thinking
about Epidemics," *Annual Review of Anthropology* 30 (2001):
363–85.

Zigas, Vincent, *Laughing Death: The Untold Story of Kuru*, Clifton
NJ: Humana Press, 1990.

Fischer, Ann, and John Lyle Fischer, "Culture and Epidemiology:
A Theoretical Investigation of Kuru," *Journal of Health and
Social Behavior* 2: (1961): 16–25.

Mathews, J. D., Robert Glasse, and Shirley Lindenbaum, "Kuru
and Cannibalism," *The Lancet* 2, no. 7565 (August 24, 1968):
449–52.

Gibbs, Clarence Joe, D. Carleton Gajdusek, D. M. Asher, et al., "Creutzfeldt-Jakob Disease (Spongiform Encephalopathy): Transmission to the Chimpanzee," *Science* 161 (July 26, 1968): 388–89.

Gajdusek, D. Carleton, "Autobiography," Nobel Foundation, 1976, http://www.nobel.se/medicine/laureates/1976/gajdusek-autobio.html.

8: Slow Virus

Dickinson, Alan G., H. Fraser, V. M. Meikle, et al., "Competition Between Different Scrapie Agents in Mice," *Nature New Biology* 237, no. 77 (1972): 244–245.

Alper, Tikvah, W. A. Cramp, D. A. Haig, et al., "Does the Agent of Scrapie Replicate Without Nucleic Acid?" *Nature* 214 (1967): 764–66.

Watson, J. D., and F. H. C. Crick, "Molecular Structure of Nucleic Acids," *Nature* 171 (1953): 737–38.

Hunter, G. D., and G. C. Millson, "Attempts to Release the Scrapie Agent from Tissue Debris," *Journal of Comparative Pathology* 77, no. 3 (1967): 301–7.

Hunter, G. D., *Scrapie and Mad Cow Disease*, New York: Vantage Press, 1993.

Griffith, J. S., "Self-Replication and Scrapie," *Nature* 215 (1967): 1043–44.

9: Prions

Prusiner, Stanley B. "Prions," Nobel Lecture, 1997: http://www.nobel.se/medicine/laureates/1997/prusiner-lecture.pdf.

Rhodes, Richard, *Deadly Feasts: Tracking the Secrets of a Terrifying New Plague,* New York: Touchstone, 1998.

Merz, Patricia A., R. A. Somerville, H. M. Wisniewski, et al.,

"Scrapie-Associated Fibrils in Creutzfeldt-Jakob Disease," *Nature* 306, no. 5942 (1983): 474–76.

Merz, Patricia A., Robert A. Somerville, H. M. Wisniewski, et al., "Abnormal Fibrils from Scrapie-Infected Brain," *Acta Neuropathologica* (Berlin), 54, no. 1 (1981): 63–74.

Somerville, Robert A., Patricia A. Merz, and R. I. Carp, "Partial Copurification of Scrapie-Associated Fibrils and Scrapie Infectivity," *Intervirology* 25, no. 1 (1986): 48–55.

Prusiner, Stanley B., "Novel Proteinaceous Infectious Particles Cause Scrapie," *Science* 216, no. 4542 (1982): 136–44.

Prusiner, Stanley B., D. F. Groth, and D. C. Bolton, "Purification and Structural Studies of a Major Scrapie Prion Protein," *Cell* 38, no. 1 (1984): 127–34.

Turk, E., D. B. Teplow, L. E. Hood, et al., "Purification and Properties of the Cellular and Scrapie Hamster Prion Proteins," *European Journal of Biochemistry* 176, no. 1 (1988): 21–30.

Lee, I. Y., D. Westerway, A. F. Smit, et al., "Complete Genomic Sequence and Analysis of the Prion Protein Gene Region from Three Mammalian Species," *Genome Research* 8, no. 10 (1998): 1022–37.

Palmer, M. S., A. J. Dryden, J. T. Hughes, et al., "Homozygous Prion Protein Genotype Predisposes to Sporadic Creutzfeldt-Jakob Disease," *Nature* 352, no. 6333 (1991): 340–42.

Mead, S., M. P. Stumpf, J. Whitfeld, et al., "Balancing Selection at the Prion Protein Gene Consistent with Prehistoric Kurulike Epidemics," *Science* 300, no. 5619 (2003): 640–43.

Tahiri-Alaoui, A., A. C. Gill, P. Disterer, et al., "Methionine 129 Variant of Human Prion Protein Oligomerizes More Rapidly Than the Valine 129 Variant: Implications for Disease Susceptibility to CJD," *Journal of Biological Chemistry,* e-pub May 6, 2004.

Hegde, R. S., J. A. Mastrianni, M. R. Scott, et al. "A Transmembrane Form of the Prion Protein in Neurodegenerative Disease," *Science* 279, no. 5352 (1998): 827–34.

Hegde, R. S., P. Tremblay, D. Groth, et al., "Transmissible and Genetic Prion Diseases Share a Common Pathway of Neurodegeneration," *Nature* 402, 6763 (1999): 822–26.

10: The Silencing

Marsh, Richard F., and R. H. Kimberlin, "Comparison of Scrapie and Transmissible Mink Encephalopathy in Hamsters. II. Clinical Signs, Pathology, and Pathogenesis," *Journal of Infectious Diseases* 131 (1975): 104–10.

Buyukmihci, N., M. Rorvik, and R. F. Marsh, "Replication of the Scrapie Agent in Ocular Neural Tissues," *Proceedings of the National Academy of Sciences, USA* 77 (1980): 1169–71.

McNair, Joel, "BSE: A Ticking Time Bomb for Downer Cows?" *Agri-View,* June 17, 1993.

Marsh, Richard F., "Transmissible Mink Encephalopathy," in *Prion Disease of Humans and Animals* (S. B. Prusiner, ed.), Chichester, England: Ellis Horwood, 1993, 299–306.

Rampton, Sheldon, and John Stauber, *Mad Cow U.S.A.: Could the Nightmare Happen Here?,* Monroe, Maine: Common Courage Press, 1997.

Raben, Maurice S., "Preparation of Growth Hormone from Pituitaries of Man and Monkey," *Science* 125, no. 3253 (May 3, 1957): 883–84.

Raben, Maurice S., "Treatment of a Pituitary Dwarf with Human Growth Hormone," *Journal of Clinical Endocrinology and Metabolism* 18, no. 8 (1958): 901–3.

Brown, Paul, "Human Growth Hormone Therapy and Creutzfeldt-Jakob Disease: A Drama in Three Acts," *Pediatrics* 81, no. 1 (1988): 85–92.

Fradkin, J. E., L. B. Schonberger, J. L. Mills, et al., "Creutzfeldt-Jakob Disease in Pituitary Growth Hormone Recipients in the United States," *Journal of the American Medical Association* 265, no. 7 (1991): 880–84.

Mills, J. L., L. B. Schonberger, D. K. Wysowski, et al., "Long-term Mortality in the United States Cohort of Pituitary-Derived Growth Hormone Recipients," *Journal of Pediatrics* 144, no. 4 (2004): 430–36.

11: Mad Cows

Hornsby, M., "Farmer Describes Horror at Seeing the Birth of BSE," *The Times* (London), June 24, 1996.

Wells, G. A. H., et al., "A Novel Progressive Spongiform Encephalopathy in Cattle," *Veterinary Record* 121 (1987): 419–20.

Martin, P., "The Mad Cow Deceit," *Night & Day* (supplement) in *Mail on Sunday*, May 12, 1996.

Wilesmith, J., "Transcript of Oral Evidence Given to the BSE Enquiry," Day 35 of the Hearing. June 22, 1998: http://www.bseinquiry.gov.uk/files/tr/tab35.pdf.

Young, J., "Mystery Disease Strikes at Cattle," *The Times* (London), December 29, 1987.

"Bovine Spongiform Encephalopathy: Minister's Meeting" on April 14, 1988." Meeting Minutes: http://www.bseinquiry.gov.uk/files/yb/1988/04/14001001.pdf.

"1988 Disease Update. Bovine Spongiform Encephalopathy," *Veterinary Record* 122 (1988): 477–78.

Brown, D., "Raging Madness Attacks Cattle," *Sunday Telegraph*, April 24, 1988.

Cahill, L., "Spongiform Fear Grows," *Farming News*, April 22, 1988.

Suich, J. C. "MAFF Briefing and Talking Points for Ministers Meeting the Press," April 21, 1988: http://www.bseinquiry.gov.uk/files/yb/1988/04/21002001.pdf.

Bradley, R., "MAFF Memo on Farming News Article," 1988: http://www.bseinquiry.gov.uk/files/yb/1988/04/22001001.pdf.

Wilesmith, J. In Confidence. BSE—A Summary of Investigations Completed and in Progress and Potential Actions for Control.

BSE Inquiry Records, May 3, 1988: http://www.bseinquiry.gov.uk/files/yb/1988/05/03001001.pdf.

12: Cover-up

"Notes of a Meeting between Representatives of UKRA, GAFTA, UKASTA and the Ministry of Agriculture to Discuss Bovine Spongiform Encephalopathy," June 2, 1988: http://www.bseinquiry.gov.uk/files/yb/1988/06/01008001.pdf.

"Cattle Herds Hit by New Disease," *The Independent*. June 2, 1988.

Holt, T. A., and J. Philips, "Bovine Spongiform Encephalopathy," *British Medical Journal* 296, no. 6636 (1988): 1581–82.

"Danger of Killer Meat in Beef Pies," *Today*, June 3, 1988.

Lawrence, A. J., "Letter to Sir Richard Southwood FRS, Expert Working Party on BSE," June 6, 1988: http://www.bseinquiry.gov.uk/files/yb/1988/06/06003001.pdf.

"Minutes of MAFF Meeting Held at Tolworth," June 13, 1988, 11:30 AM: http://www.bseinquiry.gov.uk/files/yb/1988/06/13001001.pdf.

Bradley, R., "BSE and Mink," memo of July 11, 1988.

Watson, W. A., "In Confidence. Visit to the CVL of Dr. Tim Holt," July 27, 1988: http://www.bseinquiry.gov.uk/files/yb/1988/07/27003001.pdf.

Taylor, D. M. "Letter to Dr. J. Wilesmith Regarding Lime Pits and Scrapie," August 9, 1988: http://www.bseinquiry.gov.uk/files/yb/1988/08/09003001.pdf.

Southwood, R., "Letter to E. W. Poole," September 8, 1988.

Proud, A. J. "BSE and Rendering Plants," memorandum, September 20, 1988.

"Mice Catch Cow Madness," *New Scientist*, October 8, 1988.

Prusiner, Stanley, "Letter to GAH Wells Requesting BSE Material," September 26, 1988: http://www.bseinquiry.gov.uk/files/yb/1988/09/26002001.pdf.

Fraser, H., "Letter to Gerald A.H. Wells Denying Prusiner's Request for Collaboration," October 13, 1988: http://www.bseinquiry.gov.uk/files/yb/1988/10/13002001.pdf.

"The Southwood Report," February 21, 1989: http://www. bseinquiry.gov.uk/report/volume4/chapter9.htm.

Meldrun, K. C. "BSE: Memo on Disposal of Carcasses," December 2, 1988: http://www.bseinquiry.gov.uk/files/yb/1988/12/02006001.pdf.

"BSE Cover Urged," *British Farmer*, December 1989.

Meldrum, K. C. "Memo. Bovine Spongiform Encephalopathy in the Republic of Ireland," January 24, 1989: http://www.bseinquiry.gov.uk/files/yb/1989/01/24005001.pdf.

Hornsby, M., "Meat from Diseased Cattle May Be on Sale to the Public," *The Times* (London), May 19, 1989.

Erlichman, J., "Government to Ban Use of Beef Brains in Pies," *The Guardian*, May 25, 1989.

Hornsby, M., "Ban Expected on 'Mad Cow' Organs," *The Times* (London), June 12, 1989.

Wells, G. A. H., "Perceptions of Unconventional Slow Virus Disease in the U.S.A.," June 1989: http://www.bseinquiry.gov.uk/files/yb/1989/06/21002001.pdf.

Lawrence, A. J., "BSE: Exports of Meat and Bonemeal to Other Member States," July 3, 1989: http://www.bseinquiry.gov.uk/files/yb/1989/07/03005001.pdf.

McLean, D., "Extract House of Commons Hansard," January 11, 1990: http://www.bseinquiry.gov.uk/files/yb/1990/01/11004001.pdf.

"Mad Cattle Meat Racket," *Today,* January 4, 1990.

Erlichman. J., "Scientists Confirm First Case of 'Mad Cat' Disease," *Financial Times,* May 11, 1990.

Erlichman, J., "Councils Ban Beef as BSE Fears Spread," *The Guardian*, May 16, 1990.

Heaton-Jones, K. Press Release, National Health Service, May 17, 1990.

Memo: "Professor Lacey's Evidence to the Select Committee," June 25, 1990:
    http://www.bseinquiry.gov.uk/files/yb/1990/06/25009001.pdf.

Collinge, J., F. Owen, M. Poulter, et al., "Prion Dementia Without Characteristic Pathology," *The Lancet 7*, 336: (8706), (1990): 7–9.

Wells, G. A. H., "Pathology Report," August 20, 1990:
    http://www.bseinquiry.gov.uk/files/yb/1990/08/20003001.pdf.

Clinton, J., P. L. Lantos, M. Rossor, et al., "Immunocytochemical Confirmation of Prion Protein," *The Lancet*, 336, no. 8713 (1990): 515.

Press Release: "Bovine Spongiform Encephalopathy (BSE): Studies of the Offspring of Infected Cattle," March 27, 1991:
    http://www.bseinquiry.gov.uk/files/yb/1991/03/27007001.pdf.

Dejevsky, M., and M. Hornsby, "British Mad Cow Expert Flies to Moscow," *The Times* (London), January 7, 1992.

Lowson, R., "Memo on Professor Lacey: BSE," April 22, 1992.

Shaw, D., "Keep Calm Over New Mad Cow Alert, Shoppers Told," *Evening Standard*, June 2, 1992.

13: The Tipping Point

Harding, L., "Mad Cow Meal Destroyed my Daughter's Life," *Daily Mail*, January 25, 1994.

Watkins, A., "Human Mad Cow Disease Tests on Dead Teenager," *Today*, May 24, 1995.

Bradley, R., "Possible Residual Origins of BSE Infection in Feed for Cattle," MAFF memo, July 5, 1995.

Fleetwood, A. J., "Separation and Staining of SBOs at Slaughterhouses and Head Boning Plants," results of a national survey, July 6, 1995.

"Meat Worker Gets CJD," *The Times* (London), August 17, 1995.

"CJD and the Media," Industry News Flash, MAFF, October 24, 1995.

Robb, G., "CJD: Today Newspaper. MAFF Memo," November 10, 1995.

Lacey, R., "Mad Cows and Ministries," *Today*, November 15, 1995.

Press release, Ministry of Health, "If You're in Two Minds about Serving Beef, a Chance to Digest the Facts," chief medical officer, Ministry of Health, November 19, 1995.

Wight, A., "CJD With Young Age of Onset," MAFF memo, February 15, 1996.

Rubery, E. B. "CJD or BSE," Memo, March 8, 1996.

Spongiform Encephalopathy Advisory Committee, Minutes of the 25th Meeting Held on March 8, 1996, at the Department of Health, Skipton House, London.

Hogg, Douglas, and Stephen Dorrell, "BSE and CJD," Letter to the Prime Minister, March 18, 1996: http://www.bseinquiry.gov.uk/files/yb/1996/03/18003001.pdf.

Maguire, K., "We Have Already Eaten 1,000,000 Mad Cows," *Daily Mirror*, March 20, 1996.

"Government Kept Us in the Dark, Says CJD Victim's Husband," *PA News*, March 21, 1996.

Rampton, Sheldon, and John Stauber, *Mad Cow U.S.A.: Could The Nightmare Happen Here?*, Monroe, Maine: Common Courage Press, 1997.

Southwood, R. "Could More Have Been Done Sooner?" *Daily Telegraph*, March 28, 1996.

Rhodes, Richard, *Deadly Feasts: Tracking the Secrets of a Terrifying Plague,* New York: Touchstone, 1998.

Spongiform Encephalopathy Advisory Committee; Minutes of the 81st Meeting Held on February 25, 2004. Item 21: http://www.seac.gov.uk/minutes/final81.pdf.

Kirby, L., P. Lehman, and A. Majeed, "Dementia in People Aged 65 and Over: A Growing Problem?" *Population Trends* 92 (1998): 23–28.

14: Prime Cuts

Rampton, Sheldon, and John Stauber, *Mad Cow U.S.A.: Could The Nightmare Happen Here?*, Monroe, Maine: Common Courage Press, 1997.

Marsh, Richard F., "Bovine Spongiform Encephalopathy in the United States," *Journal of the American Veterinary Medical Association* 196, no. 10 (1990): 1677.

Marsh, Richard F., "Letter," *Journal of the American Veterinary Medical Association* 197, no. 4, August 15, 1990.

Nikiforuk, Andrew, "An Issue Comes to a Head," *Globe and Mail*, January 8, 2004.

Smith, Van, "Meltdown: What Happens to Dead Animals at Baltimore's Only Rendering Plant?" *Baltimore City Paper*, September 27, 1995.

Winfrey, Oprah, "Oprah's Report on Mad Cow Disease," (show transcript), April 15, 1996:
http://www.mcspotlight.org/media/television/oprah_transcript.html.

Rampton, S., and J. Stauber, *op. cit.*

15: The Alzheimer's Nightmare

Gibbons, R. V., R. C. Holman, E. D. Belay, et al., "Creutzfeldt-Jakob Disease in the United States: 1979–1998," *Journal of the American Medical Association* 284, no. 18 (2000): 2322–23.

Kmietowicz, Z., "Surgery Increases Risk of Sporadic CJD," *British Medical Journal* 318, no. 7184 (1999): 625A.

Mehta, J. S., and W. A. Franks, "The Sclera, the Prion, and the Ophthalmologist," *British Journal of Ophthalmology* 86, no. 5 (2002): 587–92.

Frosh A., et al., Editorial, "Iatrogenic vCJD from Surgical Instruments," *British Medical Journal,* 322 (June 30, 2001): 1558–59.

Williamson, E. "Amid Mad-Cow Fear, Worries in Md. Death: Lack of Autopsies Hampers Research," *Washington Post*, January 11, 2004.

Mitchell, Steve, "Mad Cow: Linked to Thousands of CJD Cases?" UPI, December 29, 2003.

Manuelidis, E. E., and L. Manuelidis, "Suggested Links Between Different Types of Dementias: Creutzfeldt-Jakob Disease, Alzheimer Disease, and Retroviral CNS Infections," *Alzheimer Disease and Associated Disorders* 3 no. 1–2 (1989): 100–109.

Boller, F., O. L. Lopez, and J. Moossy, et al., "Diagnosis of Dementia: Clinicopathologic Correlations," *Neurology* 39 no. (1) (1989): 76–79.

Hoyert, D. L., "Mortality Trends for Alzheimer's Disease 1979–1991. National Center for Health Statistics," *Vital Health Statistics* 20, no. 28 1996.

Hebert, L. E., P. A. Scherr, J. L. Bienias, et al., "Alzheimer's Disease in the U.S. Population: Prevalence Estimates Using the 2000 Census," *Archives of Neurology* 60, no. 8 (August 2003): 1119–22.

"2000 Population and Alzheimer's Disease Prevalence Projections," Alzheimer's Association, 2002: http://search.alz.org/Media/newsreleases/archived/currenttotalschart.htm.

Stagg, Elliott V., "Alzheimer's Deaths on the Rise," *Amednews*, March 8, 2004.

Kirby, L., P. Lehmann, and A. Majeed, "Dementia in People Aged 65 and Over: A Growing Problem?" *Population Trends* 92 (1998): 23–28.

Hannaford, R., "Alzheimer's: A Disease of the Young?" *BBC Report*, November 10, 2000.

Dermaut, B., E. A. Croes, R. Rademakers, et al., "PRNP Val129 Homozygosity Increases Risk for Early-Onset Alzheimer's Disease," *Annals of Neurology* 53, no. 3 (2003): 409–12.

Riemenschneider, M., N. Klopp, W. Xiang, et al., "Prion Protein Codon 129 Polymorphism and Risk of Alzheimer Disease," *Neurology* 63, (2004): 364–66.

Hendrie, H. C., A. Ogunniyi, K. S. Hall, et al., "Incidence of Dementia and Alzheimer Disease in Two Communities: Yoruba Residing in Ibadan, Nigeria, and African Americans Residing in Indianapolis, Indiana," *Journal of the American Medical Association* 285, no. 6 (2001): 739–47.

16: Clusters

Asante, E. A., J. M. Linehan, M. Desbruslais, et al., BSE Prions Propagate as Either Variant CJD-like or Sporadic CJD-like Prion Strains in Transgenic Mice Expressing Human Prion Protein, *EMBO J* 21 (2002): 6358–66.

"BSE Linked to Further CJD Cases," *BBC News*, November 28, 2002.

Glatzel, M., Rogivue C., Ghani A., et al., "Incidence of Creutzfeldt-Jakob Disease in Switzerland," *The Lancet* 360 (2002): 139–41.

Mitchell, Steve, "Mad Cow: Linked to Thousands of CJD Cases?" UPI, December 29, 2003.

Max, D. T., "The Case of the Cherry Hill Cluster," *The New York Times Magazine*, March 28, 2004.

Usborne, D. "The Amateur Sleuth, the CJD Victims, and a Link to a Day at the Races," *The Independent*, April 7, 2004.

Pearsall, R., "CJD Cases Continue to Grow," (New Jersey) *Courier-Post*, April 6, 2004.

McNeil, D. G., "Health Officials' Inquiry Finds No Evidence of Mad Cow Disease at New Jersey Track," *The New York Times*, May 8, 2004.

Zanusso, G., S. Ferrari, F. Cardone, et al., "Detection of Pathologic Prion Protein in the Olfactory Epithelium in Sporadic Creutzfeldt-Jakob Disease," *New England Journal of Medicine* 348 (2003): 711–19.

Glatzel, M., E. Abela, M. Maissen, et al., "Extraneural Pathologic Prion Protein in Sporadic Creutzfeldt-Jakob Disease," *New England Journal of Medicine* 349 (2003), 1812–20.

Thomzig, A., W. Schulz-Schaeffer, C. Kratzel, et al., "Preclinical Deposition of Pathological Prion Protein PrP(Sc) in Muscles of Hamsters Orally Exposed to Scrapie," *Journal of Clinical Investigation* 113, no. 10 (2004): 1465–72.

Aguzzi, A. and M. Glatzel, "vCJD Tissue Distribution and Transmission by Transfusion—A Worst-Case Scenario Coming True?" *The Lancet* 363, no. 9407 (2004): 411–12.

"CJD Fears Prompt Blood Donor Ban," *BBC News*, March 16, 2004.

Hilton, D. A., A. C. Ghan., L. Conyers, et al., "Prevalence of Lymphoreticular Prion Protein Accumulation in UK Tissue Samples," *Journal of Pathology*, e-pub May 21, 2004.

Connor, S., "Scientists Fear Hidden Epidemic of vCJD," *The Independent*, May 21, 2004.

Hill, A. F., and J. Collinge, "Species-Barrier-Independent Prion Replication in Apparently Resistant Species" *APMIS* 110 (2002): 44–53.

Hill, A. F., and J. Collinge, "Subclinical Prion Infection," *Trends in Microbiology* 11, no. 12 (2003): 578–84.

17: Mad Deer

Williams, E. S., and S. Young, "Chronic Wasting Disease of Captive Mule Deer: A Spongiform Encephalopathy," *Journal of Wildlife Diseases* 16 (1980): 89–98.

Williams, E. S., M. W. Miller, and E. T. Thorne, "Chronic Wasting Disease: Implications and Challenges for Wildlife Managers," presented at the North American Wildlife and Natural Resources Conference, April 2002.

Gerhardt, G., T. Hartman, and L. Kilzer, "Killer in the Herds," *Rocky Mountain News*, June 2002.

Miller, M. W., and E. S. Williams, "Prion Disease: Horizontal Prion Transmission in Mule Deer," *Nature* 425, no. 6953 (2003): 35–36.

Spraker, T. R., M. W. Miller, E. S. Williams, et al., "Spongiform Encephalopathy in Free-Ranging Mule Deer (*Odocoileus hemionus*), White-Tailed Deer (*Odocoileus virginianus*) and Rocky Mountain Elk (*Cervus elaphus nelsoni*) in Northcentral Colorado," *Journal of Wildlife Diseases* 33, no. 1 (1997): 1–6.

"Chronic Wasting Disease Appears in Canada," *SCWDS Briefs*, 12:1, April 1996.

"Facts About Chronic Wasting Disease," South Dakota information release, July 2004.

"Chronic Wasting Disease," Nebraska Game and Parks Commission, 2002: http://www.ngpc.state.ne.us/wildlife/guides/CWD/cwd.asp.

Reed, Ollie, "Diseased Deer Outside Range Stuns Game Official," *Albuquerque Tribune*, February 15, 2003.

Van de Kamp Nohl, Mary, "The Killer Among Us: What State Officials Aren't Telling You About Chronic Wasting Disease—the Politics and Blunders Behind Its Spread and True Dangers," *Milwaukee Magazine*, December 2002.

Hamir, A. N., R. C. Cutlip, J. M. Miller, et al., "Preliminary Findings on the Experimental Transmission of Chronic Wasting Disease Agent of Mule Deer to Cattle," *Journal of Veterinary Diagnostic Investigation*, 13 (2001): 91–96.

Imrie, R., "Chronic Wasting Disease: New Testing Finds Indicators of Deer Disease in 14 New Counties," *Associated Press*, March 11, 2004.

Gould, D. H., J. L. Voss, M. W. Miller, et al., "Survey of Cattle in Northeast Colorado for Evidence of Chronic Wasting Disease: Geographical and High-Risk Targeted Sample," *Journal of Veterinary Diagnostic Investigation* 15, no. 3 (May 2003): 274–77.

18: More Tainted Meat

Gerhardt, G., T. Hartman, and L. Kilzer, "Killer in the Herds," *Rocky Mountain News*, June 2002.

Davanipour, Z., M. Alterm, E. Sobel, et al., "Transmissible Virus Dementia: Evaluation of a Zoonotic Hypothesis," *Neuroepidemiology* 5, no. 4 (1986): 194–206.

Berger, J. R., E. Waisman, and B. Weisman, et al., "Creutzfeldt-

Jakob Disease and Eating Squirrel Brains," *The Lancet* 350, no. 9078 (1997): 642.

Karun, M., and B. M. Patten, "Creutzfeldt-Jakob Disease: Possible Transmission to Humans by Consumption of Wild Animal Brains," *American Journal of Medicine* 76, 1984: 142–45.

"Fatal Degenerative Neurologic Illnesses in Men Who Participated in Wild Game Feasts—Wisconsin, 2002," *Morbidity Mortality Weekly Reports* 52, no. 7 (February, 21, 2003): 125–27.

Belay, E. D., P. Gambetti, L. B. Schonberger, et al., "Creutzfeldt-Jakob Disease in Unusually Young Patients Who Consumed Venison," *Archives of Neurolog* 58, no. 10 (2001): 1673–78.

"Chronic Wasting Disease Threatens Utah's Deer and Elk," Utah Division of Wildlife:

http://www.wildlife.utah.gov/hunting/biggame/cwd/.

Mitchell, Steve, "USDA Suspected of Hiding Mad Cow Cases," *UPI*, February 9, 2004.

Peltier, A. C., M. Pastone, P. Reading, et al., "Two Cases of Early Onset Sporadic Creutzfeldt-Jakob Disease in Michigan," presented at American Academy of Neurology Annual Meeting, Denver Colorado, 2002.

Van de Kamp Nohl, Mary, "The Killer Among Us: What State Officials Aren't Telling You About Chronic Wasting Disease-The Politics and Blunders Behind Its Spread and True Dangers," *Milwaukee Magazine*, December 2002.

Hamir, A. N., R. C. Cutlip, J. M. Miller, et al., "Preliminary Findings on the Experimental Transmission of Chronic Wasting Disease Agent of Mule Deer to Cattle," *Journal of Veterinary Diagnostic Investigation* 13 (2001): 91–96.

Imrie, R., "Chronic Wasting Disease: New Testing Finds Indicators of Deer Disease in 14 New Counties," *Associated Press*, March 11, 2004.

Sprengelmeyer, M. E., "National CWD Plan Is 'Gathering Dust': Strategy Drafted to Fight Disease Bogged Down in Red Tape," *Rocky Mountain News*, April 7, 2004.

Birmingham, K., "TSE Threat to USA Increases," *Nature Medicine* 8, no. 5 (May 2002): 431.

19: The Monitors

Onet, George E., "Animal Mutilations: What We Know," National Institute for Discovery Science, 1997: http://www.nidsci.org/articles/animal1.php.

The Meeker Herald, Meeker, Colorado, September 6, 1975.

Smith, Frederick W., *Cattle Mutilation: The Unthinkable Truth*, Cedaredge, CO: Freedland, 1976.

Kagan, Daniel, and Ian Summers, *Mute Evidence*, New York: Bantam Books, 1984.

Brown, L. M. "Covert Microbiological Experimentation and Livestock Losses," report submitted to National Institute for Discovery Science, 2001.

"Summary Report on a Wave of UFO/Helicopters and Animal Mutilations in Cascade County, Montana 1974–1977," National Institute for Discovery Science Report, February 2001: http://www.nidsci.org/pdf/wolverton_report.pdf.

Day, G. I., D. Sanford, S. D. Schernitz, et al., "Capturing and Marking Wild Animals," in *Wildlife Management Techniques Manual* (Schemnitz, S. D. ed.), 4th ed., Bethesda, MD: The Wildlife Society, Inc, 1980.

Gates, C. C., B. T. Elkin, and D. C. Dragon, "Investigation, Control and Epizootiology of Anthrax in a Geographically Isolated, Free-Roaming Population in Northern Canada," *Canadian Journal of Veterinary Research* 59, no. 4 (1995): 256–64.

Kelleher, Colm A., G. Onet, and E. Davis, "Final Report: Investigation of the Unexplained Death of a Cow in N.E. Utah, October 16, 1998," National Institute for Discovery Science, 1999: http://www.nidsci.org/articles/ucd_report1.php.

"Investigation of a Report of Animal Mutilation in Dupuyer, Montana, on June 27, 2001," National Institute for Discovery Science Report, January 3, 2002: http://www.nidsci.org/pdf/montana_cattlemutilation.pdf.

Mannaioni, G., R. Carpenedo, A. M. Pugliese, et al., "Electrophysiological Studies on Oxindole, a Neurodepressant Tryptophan Metabolite," *British Journal of Pharmacology* 125 (1998): 1751–60.

Orcutt, J. A., J. P. Prytherch, M. Konicov, et al., "Some New Compounds Exhibiting Selective CNS Depressant Activities. Part 1. Preliminary Observations," *Archives of International Pharmacodynamics* 152 (1964): 121–31.

"Investigation of a Mutilation Report in Cache County, Utah," National Institute for Discovery Science Report, July 16, 2002: http://www.nidsci.org/pdf/cache_county_mutilation.pdf.

Donovan, R. and K. Wolverton, *Mystery Stalks the Prairie*, Raynesford, MT: THAR Institute, 1976.

"Results of a Survey Among Bovine Practitioners Concerning Animal Mutilation," National Institute for Discovery Science Report, 1997: http://www.nidsci.org/articles/bovinepractioners.php.

Tuo, W., K. I. O'Rourke, D. Zhuang, et al., "Pregnancy Status and Fetal Prion Genetics Determine PrPSc Accumulation in Placentomes of Scrapie-Infected Sheep," *Proceedings of the National Academy of Sciences, USA,* 99 no. 9 (2002): 6310–11.

Tuo, W., D. Zhuang, D. P. Knowles, et al., "Prp-c and Prp-Sc at the Fetal-Maternal Interface," *Journal of Biological Chemistry*, 276, no. 21 (2001): 18229–34.

Van Keulen, L. J., M. E. Vromans, and F. G. van Zijderveld, "Early and Late Pathogenesis of Natural Scrapie Infection in Sheep," *APMIS* 110, no. 1 (2002): 23–32.

Bartz, J. C., A. E. Kincaid, and R. A. Bessen, "Rapid Prion Neuroinvasion Following Tongue Infection," *Journal of Virology* 77, no. 1 (2003): 583–91.

Head, M. W., V. Northcutt, K. Rennison, et al., "Prion Protein Accumulation in Eyes of Patients with Sporadic and Variant

Creutzfeldt-Jakob Disease," *Investigative Ophthalmology and Visual Science* 44, no. 1 (2003): 342–46.

20: Hot Zone

"BSE Investigation in Alberta," Government of Canada/Government of Alberta news release, May 20, 2003.

McArthur, M., "Corrals Quiet, Comments Loud," *Western Producer*, August 14, 2003.

"Letter: One Cow Caused $1 Billion of Harm: Our Beef Is Safe; It's Time for the Americans to Reopen the Border," *The Edmonton Journal*, June 15, 2003.

Brasher, Philip, "Grassley Demands Firings in USDA. The Iowa Senator Says That Public Trust Was Violated When Beef Imports from Canada Were Wrongly Allowed," *Des Moines Register*, May 26, 2004.

Cook, T. "Ralph Klein Suggests Alberta Farmer Should Have Covered Up Mad Cow Case," *CP Newswire*, September 16, 2003.

Nickerson, C. "Canadians Fear Impact of Mad Cow," *Boston Globe*, May 22, 2003.

"Mad Cow Quarantine Grows," *Globe and Mail*, May 22, 2003.

"Narrative Background to Canada's Assessment of and Response to the BSE Occurrence in Alberta," CFIA Report, July 2003.

"Sheep Seized from Saskatchewan Farm," *CBC News*, June 26, 2003.

Morrison, K., "CJD Death in Saskatchewan Not Linked to Animal Illness," *CBC News*, August 9, 2002.

Kagan, Daniel, and Ian Summers, *Mute Evidence*, New York: Bantam Books, 1984.

Mahoney, J., "Mad Cow Strain Hits Canada," *Globe and Mail*, August 9, 2003.

Lawlor, Alison, "Sakatchewan Man Dies of vCJD/Mad Cow," *Globe and Mail*, August 10, 2002.

Monchuk, J., "Little Change to Beef Industry Safeguards Since

Mad Cow Found 3 Months Ago," *Canadian Press,* August 20, 2003.

21: U.S. Mad Cow

Robinson, M. M., W. J. Hadlow, D. P. Knowles, et al., "Experimental Infection of Cattle with the Agents of Transmissible Mink Encephalopathy and Scrapie," *Journal of Comparative Pathology* 113 (1995): 241–51.

Pianin, E., and G. Gugliotta, "Banning Sale of 'Downer' Meat Represents a Change in Policy: Identical Measure Was Blocked in Congress Just Weeks Ago," *Washington Post*, December 31, 2003.

Dennison, M., "McLaughlin Director Says Precautions Long Overdue," *Great Falls Tribune*, January 5, 2004.

Nikiforuk, Andrew, "An Issue Comes to a Head," *Globe and Mail*, January 8, 2004.

Weise, E., "Cattle Feed Rules Unchanged: New Regulations Tied Up in Red Tape," *USA Today*, April 15, 2004.

Blakeslee, Sandra, "U.S. Moving to New Ban for Mad Cow, Officials Say," *The New York Times*, July 10, 2004.

Mitchell, Steve, "USDA Suspected of Hiding Mad Cow Cases," *UPI*, February 16, 2004.

"Mad Cow Samples Removed from Ames Laboratory," *The Ames Tribune*, March 31, 2004.

Ortitz, J., "Mad Cow Records Probed: Criminal Inquiry Will Focus on Whether Sole U.S. Case Was a 'Downer' Animal," *The Sacramento Bee*, March 4, 2004.

Mitchell, Steve, "USDA's Key Mad Cow Witness Keeps Silent," *UPI*, March 4, 2004.

Sleeth, Peter, and Andy Dworkin, "Cow's 'Downer' Status Comes into Question," *The Oregonian*, January 23, 2004.

"Cow Brain Sandwiches Still on Some Menus," *CNN*, January 17, 2004.

McNeil, D. G., "U.S. Won't Let Company Test All Its Cattle for Mad Cow," *The New York Times*, April 10, 2004.

"Editorial: USDA Goes Mad: Meat Plant Told It Can't Test Every Cow," *The Sacramento Bee*, April 14, 2004.

Talbot, Neal, "Japan Could Say Hai to Our BSE-Tested Beef," *Daily Herald Tribune*, May 21, 2004.

McNeil, D. G., "Calls for Federal Inquiry over Untested Cow," *The New York Times*, May 6, 2004.

Mitchell, Steve, "Only 3 Mad Cow Tests Done at Texas Firm," UPI, May 4, 2004.

Mitchell, Steve, "Up to 100 More Mad Cow Cases Expected," UPI, June 30, 2004.

Supervie, V., and D. Costagliola, "The Unrecognised French BSE Epidemic," *Veterinary Research* 35, no. 3 (May–June 2004): 349–62.

22: Origins

Belay, Ermias, R. A. Maddox, E. S. Williams, et al., "Chronic Wasting Disease and Potential Transmission to Humans," *Emerging Infectious Diseases* 10, no. 6 (June 2004): 977–84.

Telling, G. C., "The Mechanism of Prion Strain Propagation," *Genome Biology* 5, no. 5 (epub April 22, 2004): 222.

Vanik, D. L., K. A. Surewicz, and W. K. Surewicz, "Molecular Basis of Barriers for Interspecies Transmissibility of Mammalian Prions," *Molecular Cell* 14, no. 1 (April 9, 2004): 139–45.

Race, R., K. Meade-White, A. Raines, et al., "Subclinical Scrapie Infection in a Resistant Species: Persistence, Replication and Adaptation of Infectivity during Four Passages," *Journal of Infectious Diseases* 186, Suppl. 2 (2002): S166–70.

Mangels, J., "Breaking through the Species Firewall," *Cleveland Plain Dealer*, April 19, 2004.

Bunk, Steve, "Chronic Wasting Disease—Prion Disease in the Wild," *Public Library of Science Biology* 2, no. 4, (April 2004).

Fatzer, R., and M. Vandevelde, "Transmissible Spongiform Encephalopathies in Animals," *Wiener Medizinische Wochenschrift* 148, no. 4 (1998): 78–85.

Lupi, Omar, "Could Ectoparasites Act as Vectors for Prion Diseases?" *International Journal of Dermatology* 42, no. 6 (2003): 425–29.

Post, K., D. Riesner, V. Walldorf, et al., "Fly Larvae and Pupae as Vectors for Scrapie," *The Lancet* 354, no. 9194 (Dec. 4, 1999): 199–204.

Davanipour, Z., M. Alter, E. Sobel, et al., "Transmissible Virus Dementia: Evaluation of a Zoonotic Hypothesis," *Neuroepidemiology* 5, no. 4 (1986): 194–206.

Promed Mail: "BSE Update 2004 (04)," Archive Number 20040515.1316, May 15, 2004.

"U.S. Agriculture Secretary Expects Further Cases of BSE," *ABC National Rural News*, May 27, 2004.

Index